T0168151

Rock-Elephant

Rock-Elephant

A Story of

Friendship

and Fishing

Sam Venable

Illustrations by Tommy Smith

OUTDOOR TENNESSEE SERIES
Jim Casada, Series Editor

THE UNIVERSITY OF TENNESSEE PRESS

Knoxville

The OUTDOOR TENNESSEE SERIES covers a wide range of topics of interest to the general reader, including titles on the flora and fauna, the varied recreational activities, and the rich history of outdoor Tennessee. With a keen appreciation of the importance of protecting our state's natural resources and beauty, the University of Tennessee Press intends the series to emphasize environmental awareness and conservation.

Copyright © 2002 by The University of Tennessee Press / Knoxville. All Rights Reserved. Manufactured in the United States of America. Paper: 1st printing, 2002; 2nd printing, 2004.

The lure image that appears throughout the book is based on a crankbait lure that was handmade by Ray Hubbard.

The paper used in this book meets the minimum requirements of ANSI/NISO Z39.48-1992 (R 1997) (Permanence of Paper). The binding materials have been chosen for strength and durability.

LIBRARY OF CONGRESS CATALOGING-IN-PUBLICATION DATA

Venable, Sam.
Rock-elephant: a story of friendship and fishing/Sam Venable; illustrations by Tommy Smith.
    p. cm.—(Outdoor Tennessee series)
ISBN 1-57233-153-4 (pbk.: alk. paper)
1. Fishing—Tennessee—Anecdotes.
2. Friendship—Tennessee—Anecdotes.
3. Hubbard, Ray, 1928–1999.
4. Venable, Sam.
I. Title.
II. Series.

SH549 .V46 2002
799'.1'1'09768—dc21   2001004260

For fishing

buddies everywhere

Other Books by Sam Venable

Photo by Paul Efird.

An Island Unto Itself

*A Handful of Thumbs
and Two Left Feet*

*Two or Three Degrees
Off Plumb*

*One Size Fits All and Other
Holiday Myths*

From Ridgetops to Riverbottoms
A Celebration of the Outdoor Life in Tennessee

I'd Rather Be Ugly Than Stuppid

Mountain Hands
A Portrait of Southern Appalachia

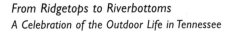

Contents

Editor's Foreword   ix
    Jim Casada, Series Editor
Foreword   xv
    Bob Hodge, Outdoor Editor,
    Knoxville News-Sentinel
Acknowledgments   xix

Saying Goodbye   2
Picture Perfect   8
Through the Bedroom   14
"Rock-Elephant"   24
Operation Code Name   32
Up Close and Personal   42
Made from Scratch   52
"Murph," a.k.a. "Queen Mary"   62
Barney and His Bullet   74
Little White Lies   80
Getting to the Point   88
Ol' Redemption   98
Horse Trading 101   108
Slop Holes   120
Mesmerizing   130
The Wailing Wall   142
Rivers in the Sky   154
For the Record   160
Solo   174

Editor's Foreword

Friendships forged in an outdoor setting often prove truly special, and they have figured prominently in some of this country's finest sporting literature. The part-fictional, part-autobiographical link between an adolescent and his grandfather described by North Carolina's Robert Ruark in *The Old Man and the Boy* is arguably the finest work of its genre ever produced in America. The partnership between the youthful Ruark and his maternal grandfather, who knew "pretty well everything," was a timeless one of the sort familiar to anyone fortunate enough to have had close contact with a relative in an outdoor setting. What made it unique was Ruark's subsequent ability to capture in prose the nostalgic wonder of those elusive days of boyhood.

Similarly, two other giants of southern sporting letters, South Carolina's Archibald Rutledge and Tennessee's Nash Buckingham, devoted some of their finest work to friendships revolving around activities afield. In a pointed reminder that the racism of yesteryear so often ascribed to affluent southerners had another, quite different side, the friendships they portray were with African Americans. Rutledge's "comrade to my heart" who brought him "a peace the world could not give" was Prince Alston. A "black hunterman" without peer, Alston was one of the many individuals Rutledge portrayed in *God's Children*. Prince also figured prominently in dozens of the stories which can be found in Rutledge's minor classics such as *Days Off in Dixie, Home By the River, An American Hunter, Those Were the Days,* and *Tom and I on the Old Plantation.*

Buckingham's erstwhile sidekick in countless joyful days spent waterfowling or following wide-ranging pointers searching for bevies of quail was the inimitable Horace Miller. Buckingham had a rare knack with dialect (as did, for that matter, Rutledge), and the way he captures Horace's enthusiasm and telling descriptions gave us a sportsman for the ages. One of the best-known of all outdoor stories is Buckingham's "De Shootinest Gent'man," and the title comes directly from Horace.

These are but a sampling of the enduring friendships to be found in outdoor literature, and it is worth noting that all of them come from southern writers. In the present work, Sam Venable joins their ranks with a compelling chronicle of a most unlikely friendship formed through fishing. I use the word "unlikely" advisedly because, at first blush, Venable and the individual about whom *Rock-Elephant* is written seem singularly suited to remind us of Rudyard Kipling's line: "The East is East and the West is West, and never the twain shall meet."

On the one hand you have a hardened, sometimes profane, and always assertive journalist—Venable. He is a man of intellect who has seen and read enough of the world for skepticism to become almost second nature with him. These characteristics and outlook on life contrast remarkably with the soft-spoken optimism of the deeply religious, relatively uneducated country preacher who collected fishing lures and fished (for souls and bass) with an abiding passion—Ray Hubbard. Plagued by all sorts of physical ailments, Hubbard was a man who exhibited, beneath a sometimes crusty exterior, an extraordinary capacity for friendship.

He and Venable became the fastest of friends, and long before Hubbard's death the journalist began to ask himself: "What am I going to do when that old man dies?" As Venable's colleague, Bob Hodge, tells us in a foreword that carries hints of the poignancy which is to follow, he found the answer to that painful, oft-asked question. This book is the answer, and I can honestly say it is one of the most moving pieces of outdoor material I have ever read.

Anyone who has read and reread Corey Ford's "The Road to Tinkhamtown," leaving the pages misty-eyed every time, will find the same sort of emotional depths here. In fact, if you read this book "kivver to kivver," as my grandfather used to put it, and don't cry, some serious soul searching is definitely in order. Yet for all the pain and sorrow punctuating these pages, there is plenty of laughter and joy as well.

Venable begins at the end—the funeral service for Reverend Ray. He was to deliver the eulogy, and this seasoned (some might even say "hard") veteran of the newsroom finds himself "scared tee-totally to death." Then, as he agonizes about how he will make it through an emotional tribute to his angling friend, a glance at the program for the service of remembrance shows him the words "Sam Venable, clergyman." The irony proves irresistible, just the tonic the author needs to eulogize his friend. The chapter concludes with a fond look backward, "when spring lasted half of the year, [when] the bass were always biting, . . . [and when] the livin' was easy."

Reminiscing on those bygone days, through good times and then ones which were increasingly bad, forms the heart of this book. It is vintage Venable in every sense. Those already familiar with Sam's writing will readily concur. As Bob Hodge notes, in east Tennessee there have long been "two types of outdoor writers: Sam Venable and everybody else." Whether old acquaintances or newcomers to the work of this splendid scribe, those who read what follows will find all of the ingredients which make Sam such a treasure—wit and whimsy, warmth and wisdom. Most of all though, this book is a tribute to human goodness as personified in Reverend Ray.

Given Ray Hubbard's humble roots and simple life, he probably never read Izaak Walton's classic, *The Compleat*

Angler. For that matter, I would venture to say that the same holds true for ninety-nine out of a hundred serious fishermen, no matter what their educational background or status in life. Yet he assuredly was "a Brother of the Angle," and he would have heartily agreed with Walton's suggestion that angling "deserves commendations; . . . it is an art worthy of the knowledge and practice of a wise man." Similarly, he would have cherished Walton's suggestion that "you will find angling to be like the virtue of humility, which has a calmness of spirit and a world of other blessings attending upon it." In effect, after his own simple fashion, that was part of the wisdom he imparted to Venable every time they shared a boat.

Their adventures began in the early 1970s, and from the outset Venable had him pegged—"Uncle Jessie Duke, Junior Samples, and P. T. Barnum, all rolled into one." Problem was, Hubbard read Sam just as well, and from the beginning took great delight in poking, prodding, and generally trying to get under his skin. As these pages will reveal, beneath the minister's simple exterior were unfathomable depths. His deep religious feelings combined with uncanny fishing ability and rare understanding of human nature. All of these things he shared with Venable, in a straightforward, unpretentious fashion.

Ray was a topwater genius, and anyone who has done much bass fishing will readily tell you that catching fish on the surface is the ultimate in bassin'—it's the icing on the angler's cake, the cherry atop his sundae. He was a masterful caster, a designer and producer of unlikely but highly effective lures (he was making and fishing buzzbaits long before the word became a standard part of the bass fisherman's parlance), and a free-flowing fountain of fishing knowledge.

The latter he shared, sometimes a bit begrudgingly and even occasionally with some downright condescension. Yet Venable would be the first to acknowledge, as he does repeatedly in these pages, that the reverend's superior attitude on all matters associated with angling was richly deserved. He could, and consistently did, fish rings around the sports writer.

Fishing forms the glue which holds the subject matter of this book together, but in many senses it isn't just a book about fishing. It is a work about the celebration of life, love of the good earth and the water which is its life's blood (and home to fish), and deep appreciation for the supreme being Reverend Ray worshiped and served. You will learn something about fishing here, although this book is far removed from the drab, dull "how-to and where-to" chronicles which form a fixation for today's magazine editor. You will also be privileged to enjoy, vicariously, an unlikely but wonderful friendship.

I'll resist the temptation to explain the book's strange but appropriate title. Likewise, when it comes to other revelations about *Rock-Elephant*, I think it best to adhere to the advice of my turkey-hunting mentor, who over the years has regularly reminded me that "a man's got to have some secrets." Rest assured, Reverend Ray had his share of secrets, and over time he shared them with the man who was both his protégé and a sort of ongoing project in inimitable fashion. By doing so he provided the raw material from which Sam Venable has shaped and polished a monument to cherished memories. The pages which follow will tug at your heartstrings, and as they do so you will be reminded repeatedly that often life's best things are simple in nature. This saga of a fishing friendship, at once lighthearted and serious, will brighten your outdoor days and lighten your armchair ways. You will, after reading it, know that you have been blessed through Venable's introducing you to one of God's good and gentle creatures, a man for whom fishing was a metaphor for life. As its general editor, I feel genuinely privileged to welcome this work to the Outdoor Tennessee series.

Jim Casada
Series Editor

Foreword

The first time I met Ray Hubbard was at an antiques and collectibles show at the Tennessee Valley Fairgrounds. He was breaking the sad news to the owner of a decades-old fishing plug that it wasn't worth very much money. Granted, the lure—a Creek Chub Pikey—is at the top of a lot of collectors' lists. But this Pikey had been through the bass wars too many times, been thrown against too many rocks, been neglected too long in some old tackle box. Ray was saying the lure's dollar value was next to nothing.

But it's not always the sound of a cash register that gets peoples' attention.

"Whose plug was this?" Ray asked the owner, a man whose salt and pepper hair and just-beginning-to-show crow's feet around the eyes would make him to be around forty-five years old.

"It was my granddad's," the man said. "He was still using it some when he took me fishing when I was a kid. I remember him using it at a farm pond out on Emory Road. I can't remember if he ever caught anything on it, but back then it was still shiny and in its box. I put it in my tackle box when I was in high school but didn't take very good care of it. I really never thought about a lure being worth much."

Ray and the man talked for a few more minutes about fishing, about grandfathers, and about collecting lures. It turned out the man was a lawyer. He decided he would clean the Pikey the best he could and put it on display in his office.

Ray told him it would still catch bass.

"No," the man said as he left, "I would hate to lose it."

The next person in line had a lure that was just as worthless. And before long, she was telling Ray about fishing with her father.

This was in 1996, and Ray's health wasn't good. Yet he stayed at the antique show for a couple more hours, identifying a few real collectable pieces, but mostly telling people that the old stuff in Dad's tackle box was just that—old stuff. Ray leaned against the corner of a glass counter and talked to dozens of people about fishing and grandfathers and collecting lures.

All of them had walked into the show with dollar signs in their eyes. All of them walked out with smiles on their faces.

Ray Hubbard could make you smile. He could make you smile with an endless stream of jokes. Not the, "Hey, have you heard this one?" kind of jokes, though. They were the kind that were being played on you when you didn't know it.

After watching Ray interact with the people for awhile, I introduced myself and said I would like to do a story about antique lures. He said he read the Knoxville newspaper and asked if I knew Sam Venable. I told him I did and asked him if he'd ever met Sam, who was once the *News-Sentinel's* outdoor editor.

"I met him once or twice, but I think you're a lot better writer than he was," Ray said. "I doubt he would remember me, but tell him I said hello."

Two days later, I walked into Sam's office and delivered the message. I told him about some "nice old guy" whom he had known years ago. I proudly told him the old guy liked my writing better than his. (In east Tennessee, there are two types of outdoor writers: Sam Venable and everybody else.) Sam just laughed. Little did I know that he and Ray had been close friends for years.

But to say they were "just friends" really doesn't do justice to the relationship. Bass may have brought them together decades earlier, but fishing had become a prop for them—a reason for Sam to get away from the office and for Ray to put kidney dialysis and heart problems and other physical ailments out of his mind.

Shortly after I met Ray, I was lucky enough to go with him and Sam to some of their secret spots. The fishing was lousy that day, but it was obvious to me that catching bass wasn't the reason they were together. Ray's poor health made it hard for him to get around some of the mudflats and creek bottoms, so we walked slowly and laughed a lot. Sam would dig his spurs into Ray mercilessly. At first, I even thought it bordered on cruelty. Then I watched as Ray would laugh and turn the spurs back on Sam, and I realized these two lived to devil each other.

I would go fishing with Ray and Sam again, I would eat lunch with them, and I would hang around with them. But I was the third wheel. Ray would get sicker, and Sam would gig him all the more. Ray would laugh that much harder. If laughter is the best medicine, Ray should have been cured ten-fold.

The last time I saw Ray was the fall before he died. He gave me a spinning rod and reel to replace one that had been stolen from my truck when he and I were fishing several months earlier. My stolen rod was run-of-the-mill discount store stuff; what Ray handed me was a collectable reel outfitted with a rod that would cost $100 if you could still buy them.

That day, Sam said, "What am I going to do when that old man dies?" at least half a dozen times. He wasn't saying it to me. He was saying it to himself.

When Ray did die, Sam spoke at the funeral but didn't say much more about it. Months later, when their beloved "slop holes" had been exposed in the mudflats, Sam packed up and went fishing alone. He didn't say anything to me or anyone else.

Later, he told me he had caught some bass and thought a lot about Ray, but I could tell there was more to it than that. He wasn't letting go of Ray. He was reaching out and grabbing hold of him even tighter.

He was remembering the jokes, the laughs, the fish, and the man. He was looking for that same stuff the lawyer found when he talked to Ray about his granddad's old Pikey. And he was still looking for the answer to that question—"What am I going to do when that old man dies?"

Sam found the answer. He wrote this book.

Bob Hodge, Outdoor Editor
Knoxville News-Sentinel
Knoxville, Tennessee
August 24, 2000

Acknowledgments

I am deeply grateful to the family of the Rev. Ray Hubbard—especially his wife Mary Jo, oldest son Major, and oldest grandson Patrick—for allowing me the opportunity to share a special part of Ray's life with readers and for their help in gathering information for this book. We all had been friends for years but learned to lean on each other emotionally in the difficult weeks and months following Ray's death. They will forever be part of my extended family and I of theirs.

No mention of family would be complete without acknowledging the support and encouragement of my wife. Mary Ann not only has endured the often-chaotic life of a newspaper-fishing-hunting widow for more than thirty years but also serves as technical specialist nonpareil, patiently showing my typewriter-trained fingers how to use a computerized word processor. Additionally, my heartfelt appreciation is expressed to Tommy Smith, who drew the superb illustrations for this text and doubles as the best son-in-law I've ever had.

Also, I thank Bob Hodge, outdoor editor of the *Knoxville News-Sentinel*, for his kind words of introduction to this text. *News-Sentinel* colleagues Lowell Branham and Susan Alexander are thanked for reading the manuscript and making many helpful editing suggestions.

Furthermore, I am grateful to the University of Tennessee Press—especially Jennifer Siler, director, and Jim Casada, editor of the "Outdoor Tennessee" series—for their assistance in this project.

Ray Hubbard was born August 14, 1928, on a farm in Jefferson County, Tennessee. He died March 25, 1999, at his home in Louisville, Tennessee, and is buried at Louisville Cemetery. A sewing machine salesman and repairman by trade, he was a Christian minister by calling. He was ordained at the Lone Oak Independent Church in 1951 and pastored there until 1957. He also served at First Wesleyan Church, 1952–74, and at Lost Creek Church, 1983–88.

I never heard Ray preach—at least not from a pulpit. Throughout the twenty-seven years we fished together, however, he launched any number of messages my way. We enjoyed each other's company immensely, teased each other incessantly, hoodwinked each other mercilessly, and aided each other professionally. Ray's name and photograph appeared under my byline in more than two dozen newspaper and magazine articles. He never flat-out said so, but I have a sneaking suspicion I was the subject of more than one sermon.

God bless you, Ray Hubbard. Save me a seat in the boat.

Sam Venable
February 6, 2001

Saying Goodbye

"This is when a man ought to be fishing."

That's how the Rev. Ray Hubbard always felt about spring.

Oh, he might have to postpone the trip long enough to work on a sermon or visit a sick friend in the hospital. He might even sacrifice a bit of fishing time to prepare his garden, too. Or maybe—if no grandsons were handy for the task—clean out the gutters and rake up the last of winter's leaves. But when the redbuds and jonquils were in bloom and the robins heralded each dawn and dusk with cheerful song, it well-nigh took divine intervention to keep him from squeezing in at least a few hours on the water.

Yet for Ray, "spring" didn't necessarily start on the vernal equinox, nor did it end at the summer solstice. Heaven forbid. Spring was too grand and glorious a season to be confined to such narrow parameters. If Ray's spring happened to coincide with the more conventional season of the year, the one everybody else celebrated, fine. He just didn't need a date book to verify the process.

Ray's spring—specifically, his fishing spring—began about the time the rest of the world was buying roses and Valentine's cards. It didn't end until the hot, humid dog days of August had settled like smog upon the land. That, in the Hubbard theory of seasonal change, was when summer officially began. And he loved it. Never once did I hear Ray complain about the heat. He could get chilled to the marrow on a frosty October morning and insist the second Ice Age had begun. But too hot for him? Never.

For Ray, fall kicked in, oh, somewhere around the middle of September. It may have caused him to bundle up, but it didn't slow down his fishing. But then came winter—arrgh!—which had to be endured for four or five long weeks until February rolled around and heralded a new beginning.

Yet here it was sure-nuff spring. Even said so on the calendar. Throughout the drive from home, I noticed the first wisps of green lace starting to emerge on the maples and willows. Already, some suburbanites were manicuring their lawns.

It was just before sunset when I reached my destination. As I opened the car door and stepped onto the asphalt parking lot, the first sound that filled my ears was a chorus of robin music. The air was cool, yet it smelled fresh and sweet.

Yes, indeed. This was the time of year a man ought to be fishing. Ray and I had spent any number of days just like this one on the rivers, lakes, and streams of eastern Tennessee. Lord only knows how many.

But not today. And never again.

I took a deep breath, said a silent prayer for confidence and guidance, and pulled open the front door of Miller Funeral Home. In a little over an hour, I would be delivering the eulogy for one of the dearest friends I'd ever had in this world.

I was scared tee-totally to death.

Major Hubbard, Ray's oldest son, met me at the entrance to the chapel. Tears glistened in his eyes. Immediately my own eyes began to burn, just as they had repeatedly over the last forty-eight hours, ever since Ray's grandson, Patrick, had telephoned in the middle of the night with the news I had been dreading for weeks.

"Can you do it, big guy?" Major said in little more than a hoarse whisper.

"I dunno," I replied, drawing in another deep breath. "I reckon if you can, I can."

"It's what Daddy wanted," he said. "You read it in his last papers just like I did. Me to sing and you to speak."

"Oh, I know," I sighed. "I just didn't realize how hard this was gonna be."

Major and I walked into the pastor's study at the side of the parlor. Dr. Charles Bailey and the Rev. Carl White, the officiating preachers and Ray's longtime friends in the ministry, greeted us with handshakes and hugs. I sat down in a chair, and we made small talk for a few moments about the order of the service. Then it finally dawned on me to look at the remembrance program someone had stuck into my hands when I signed the register.

Down toward the bottom, under the heading of "Officiating Clergymen," was Dr. Bailey's name. And the Rev. White's name. And my name.

That's when it hit me.

Despite the sadness that permeated the room, despite the solemnity of the occasion, despite the fact this was one of the darkest hours in my life, I couldn't help but chuckle.

I nudged Major with an elbow and pointed to the line. The trademark Hubbard smile widened across his face when he saw it. Before Major could speak, I said aloud what I knew he was thinking.

"The ol' rascal finally got me into the amen corner, didn't he?"

Sam Venable, clergyman.

If there was ever a doubt that the Lord works in mysterious ways, this one phrase dispelled it.

Sam Venable, clergyman.

I kept rolling the notion around in my mind. And chuckling under my breath. Every time the image of those words formed in my head, my countenance brightened. Ray Hubbard had come through for me again. When the organ music faded and the funeral director fetched us for the service, there was no doubt I could speak clearly, boldly, lovingly about my friend of twenty-seven years.

We processed single-file into the pulpit area. Off to the side, the Glory Road Quartet sang "When They Ring Those Golden Bells." After that, Major rose and filled the chapel with his rich, deep intonation of "Thank You," by Ray Boltz. Then one of the preachers nodded at me. I walked to the lectern to read the words I had written. The same words that, until a few moments before, I was convinced would stick like peanut butter in the back of my throat.

Not this time, they wouldn't. No way. The very thought of Ray up there in heaven, kicked back with a glass of iced tea and watching me squirm, was the perfect tonic for what ailed me.

Fishing and religion were integral parts of Ray's life, I told the congregation. Ray always figured both were divine callings, for when Jesus was on earth, he surrounded himself with fishermen.

I told them there's a passage in the twenty-first chapter of the book of John that reminded me of Ray every time I read it. This was after Jesus' resurrection, when he revealed himself to the disciples at the Sea of Tiberius. The men had been fishing but had not been successful. On Jesus' instruction, they cast to the other side of the boat and filled their nets.

That's usually where the story ends.

But if you'll read further down, to the eleventh verse of the chapter, you'll see where someone actually counted the number of fish in the net. The text does not generalize with vague words like a "large number of fish." It doesn't say "an exceedingly good catch," either. Instead, it points out that there were precisely one hundred fifty-three of them.

You know it had to take a real fisherman to sit down and sift through that mess and make the official tabulation.

I'm certain Ray volunteered for a job like that the minute he was processed through the Pearly Gates. What's more, he surely convinced everyone within earshot that all the fish were caught on his homemade plugs—but, unfortunately, he only had one in his tackle box.

That line brought a laugh from the congregation. Perhaps it's not the proper response to elicit during a funeral service, but I couldn't help myself. Neither could the assemblage of mourners, for many of them had shared a boat with Ray, too.

Make no mistake; I was suffering that evening. Mightily. The pain of losing so close a friend was as raw and sharp as if my face had been scrubbed with 150-grit sandpaper. But it paled in comparison to the pain I knew Ray had been enduring.

You name the disability, and Ray suffered from it. He had rheumatic fever as a child. Throughout adulthood, he fought a never-ending series of maladies that made Job's woes look like the measles. From a physical sense, the last couple of years of Ray's life truly had been hell on earth.

Heart disease. A "minor" stroke—if such a devastating blow can be relegated to degrees of severity. Arthritis. Colon cancer. Complete kidney failure and thrice-weekly sessions on dialysis.

"I'm dyin' on the installment plan" is the way he used to phrase it, not necessarily in jest. "I don't need all of this pain. Why don't you share some of it with me?"

Thankfully, Ray's suffering was over. Mine was just beginning. As dearly as I enjoyed the sport of fishing, I sensed, selfishly, that the very act of making a cast without Ray's company would likely be excruciating. In the months that followed, this hunch proved ever-so-bitterly correct.

But back in the good ol' days, back before needles began to rule his every waking hour, back when he and I would meet at the nearest launch ramp on thirty minutes' notice, back when spring lasted half of the year and the bass were always biting, you better believe the livin' was easy.

And, brother, did we ever have a time of it.

Picture
Perfect

H is photographs began arriving in March 1972, two years after I had been hired as outdoor editor of the *Knoxville News Sentinel*.

There was always one picture per envelope. They didn't come often, and each included only the sketchiest of details. Consistently, a human subject was missing. The lone image within the margins would be of one or more largemouth bass of wall-hanger proportions—six pounds, seven pounds, a few even larger—sometimes in focus, most times not. Plus two words scribbled in ink: "top-water" and "Hubbard."

I tossed every one into the trash.

Although it's not the modern practice, newspaper sports sections in those days usually were peppered with photos of proud anglers and their catches. These were marvelous fillers for the odd space here and there between stories, and there was never a shortage from which the outdoor editor could select. Boat dock operators with (a) the most basic grasp of the term "free advertising" and (b) a Polaroid camera, mailed them in by the dozens every week.

Approximately 98 percent met the same fate as the "top-water Hubbards."

The *News-Sentinel's* photojournalism standards of that era may not have been in the same league as *National Geographic*, but I had a pretty good idea of what would pass muster among readers—not to mention the sports editor. Toothless grins, snuff-stained jaws, cigarettes dangling from lips, sunburned beer bellies, blood and guts, back-lit subjects with no detail in the foreground, and those infernal shove-the-fish-close-to-the-camera-so-it-looks-bigger specials never got so much as a second glance—even though they continued to arrive, one fat envelope after another.

But by the time the fourth or fifth "top-water Hubbard" showed up, my curiosity was sufficiently piqued. If for no other reason, I wanted to know why this Hubbard character, whoever he might be, never included himself in the photo. Ever since Og and Ig were catching fish with crude bone hooks and then rendering artistic proof of the results on the wall of their cave, it has been traditional for the conqueror to hoist the victim aloft and smile approvingly. All this fellow did was hang his stringered bass from a nail, snap a shot, and scribble down the weights and dates.

Was this some kind of hoax being perpetrated on the kid who edited the outdoor page?

Did this guy keep a few frozen fish in inventory and thaw them as needed?

By now it was late June, and the blast furnaces of summer were humming. This was the time of year when bass supposedly vacated the shallows and moved into deep water. Anglers switched from day to night to pursue them. Did this goof really expect me to believe he was catching lunker largemouths, right now, on top-water lures?

I turned the photo over. There was a telephone number. No name. Just a number. Gotta be a joke among some redneck fishing buddies, I reasoned. Then again. . . .

Bristling with skepticism, I dialed the number. The sooner I dismiss this prankster, I resolved to myself as the phone on the other end began to ring, the better.

A woman's voice answered. "Singer Sewing Center," she said.

I'd love to have a video recording of my reaction to that greeting. Surely I did a visible double take.

Sewing machines?

"Ma'am, do you know anybody named 'Hubbard'?" I asked.

"Sure," she replied. "Hold on a minute."

The woman set the receiver aside. I could hear her heels clicking, quieter with each step, as she moved further along behind the counter. And then faintly, in the distance: "Raaaay! Telephone!"

More steps. Progressively louder this time.

Up came the receiver. And for the first of what would surely be ten thousand renditions over the next quarter of a century, I heard a low, deep, slow, "Hu-low."

I identified myself. Then I asked, "Is your name Ray? All the pictures you've been sending me say is 'Hubbard.'"

"That's me," he answered.

"And you've been catching these bass on top-water?"

"That's mainly all I ever fish."

"Whereabouts?"

"Oh, different places."

Riiiight, I thought. This guy's full of crap. He's stringing me along, for sure. Why am I even wasting my time?

"You'onna go with me sometime?" he abruptly volunteered.

Hmmm. Why not? That would at least put a stop to the "top-water Hubbard" pictures. And if, by some stretch of reason, he *was* for real, it might make an interesting story.

"When?" I asked.

"Tuesdays are my day off."

How odd, I thought again. I didn't know another person in the world that had Tuesdays off. Saturday, the traditional leisure break for outdoor activities, was an on-the-clock day for me. There was always a fishing feature to track down or a boat race to cover or a field trial to check out. Since it fell after the weekend rush and before the start of the next weekend's flurry, Tuesday was a good day for me to knock off.

"I'm gonna rearrange my schedule next week and work on Tuesday," I told him.

Ray's voice sounded quizzical. "If you do that, how're we gonna go fishin'?"

"We'll go because I'll be working," I responded.

He laughed. It was a melodic, staccato laugh that came through the receiver like the whinny of a horse. *"Herr-herr-herr-herr!"* I'd never heard anything quite like it, certainly not from another human.

"You mean t'tell me you really go fishin' and call it *work?*" he exclaimed.

"That's what they pay me for."

"Herr-herr-herr-herr!" Louder this time.

"Well how do I sign up?" he said. "That sure would be better than fixin' sewin' machines for a livin'."

I'd heard that kind of lame comment before. Approximately ten times a day, in fact. All outdoor writers hear it—and there's no sense getting offended, inside the newspaper business or out. Nobody thinks twice if a reporter wearing a suit and necktie sits down with some CEO over a pricey lunch—on expenses, naturally—and conducts an interview. But let another reporter put on blue jeans and a T-shirt and meet some good ol' boy at a boat dock, spend the next eight or ten hours fishing in the sun or rain, making notes and taking photographs, and it's supposed to be *work?*

Yeah, it sure is. The most fun type of work ever invented. The Peace Corps' motto notwithstanding, outdoor writing is the toughest job anybody could ever love, bar none.

"Where you wanna go?" I asked.

"You know where the Singleton launch ramp is over on Fort Loudoun Lake?" said Ray.

"Yep."

"Meet me there at daybreak."

"Daybreak? Aren't you fishin' at night this time of year?"

Ray ripped off another machine gun laugh.

"Now, why would a man want t'ruin a good night's sleep by fishin'?" he said. "Them ol' bass like to sleep, too. I'll show you how to wake 'em up."

"My boat or yours?"

"We'll take mine," he replied. "You just bring a castin' rod with some heavy line."

"How heavy?" I wanted to know. "Ten-pound? Twelve?"

Ray laughed once more. Louder than ever. I know it sounds trite to use the adjective "infectious," but that's exactly the response his laughter elicited. It was involuntary. I had to laugh back.

"Ten-pound test? Boy, you must let a lotta fish get away! You better have somethin' like seventeen or twenty. We ain't gonna be dealin' with no crumb-snatchers."

"All right. I'll see you then." And the deal was sealed.

I went out the next day and bought a spool of twenty-pound-test monofilament. The night before our trip, I fed it onto a red Ambassadeur 5000 casting reel. It felt thick, stiff, coarse, as it snaked through my fingers.

This isn't fishing line, I said to myself. It's ski rope. I'm going fishing tomorrow with a man who works with sewing thread and fishes with ski rope.

It oughta make for an interesting day.

Through
the Bedroom

I n what would become a ritual in the future, Ray was waiting when I arrived. His boat was already launched, his gear already stored on board. He glanced at his wristwatch as I wheeled into the parking lot.

He was a larger man than I expected, as if one can form a mental image with any degree of accuracy on the basis of one telephone conversation. Just shy of six feet tall. Square-shouldered, but slightly stooped. He had a shock of dark hair with no sign of gray, even though he obviously was older—a full eighteen years my senior, I would discover later on. His arms were tanned leathery brown from the summer sun. The sun probably bore some responsibility for the crow's feet etched in the corners of his eyes, but a more likely culprit was the radiant smile that beamed across his face.

Some people—confidence men and politicians come immediately to mind—never seem to develop the skill for spontaneous smiling. It always looks fake, forced. Not the case here. Ray's mouth opened into a toothy grin as natural as the split down the side of a ripe watermelon. There was just *baaarely* a hint of impishness.

He stuck out a huge mitt of a hand, radiating with sausage-thick fingers.

"You must be the late Sam Venable," he teased.

"Late?" I fired back flippantly as we shook. "You said to be here just after daylight. I had to use my headlights most of the way."

"Why, the sun's an hour high already," he admonished, "but there ain't no sense arguin'. C'mon. Let's get goin'."

In point of fact, it was not yet legal sunrise. I suppose I could have produced a time chart from the glove compartment of my truck for proof, but it wasn't necessary. If you're raised in the South, you can spot good ol' boy jousting half a mile away. It's a Y-chromosome thing. You either understand it or you don't.

"D'juget heavy line?" he asked.

"Yeah," I said, reaching for the casting rod and reel I had spooled with ski rope. "How's this?"

Ray rubbed the line between his thumb and index finger. His lips pursed. His face wrinkled in mock thought.

"A little light," he finally pronounced, "but you might be able to get by with it."

I reached for a lighter casting rod. And a spinning rod. And the first of two large tackle boxes. Two years of on-the-job training had taught me to be prepared.

Ray's mouth fell open in animated surprise.

"Whad'you do? Bring a whole sportin' goods store?"

Then he answered his own question—again, as incorrectly and inconsequentially as the debate over sunrise.

"Why, no. You sportswriters don't never go to no stores. Those manufacturers send you all that stuff so you'll write about it. Why, I bet you could stock a store from your garage."

Ray pointed to one of my tackle boxes, which, in truth, was about the size of an overnight suitcase.

"Just leave that ol' possum belly in your truck," he said. "I've got everything we'll need to fish with."

Then he stopped short.

"No, on second thought, bring it with you. I might want to poke around in it and see if there's somethin' in there you don't need as bad as you think you do."

Amazing. Utterly amazing. I had known this guy for what—two minutes?—and he was already giving me grief. What's weird is, I enjoyed every moment of the show. Here was Uncle Jessie Duke, Junior Samples, and P.T. Barnum, all rolled into one.

Ray helped me gather my gear, complaining that his back would surely go out in the fifty feet between the parking lot and the lakeshore because of the unnecessary load he was being forced to bear.

We walked to his boat, a fiberglass fourteen-footer, stripped of everything but the essentials. It was an open boat with plank seats fore and aft. An eighteen-horsepower outboard motor was bolted to the transom. Next to it sat a small electric motor. Ray's fishing gear was equally Spartan: a six-and-one-half-foot bait-casting rod that looked wicked enough to double as a push pole, a "tackle box" that had held cigars in its former life, and a gold, patent-leather woman's purse.

"What's that for?" I asked. "You expecting company?"

Ray reached for the purse, popped the snap, and held it open. Inside were five or six blue plastic worms.

"Got this for a dime at a flea market," he replied. "Best thing there ever was for carryin' worms. Keeps 'em from gettin' wrinkled."

Tournament bass fishing was still in its relative infancy in the early 1970s, but already the reservoirs of the South were dotted with a new, sleek, fast type of vessel known, blandly and descriptively, as the "bass boat." Except there wasn't, nor isn't, anything bland about them. You could have bought a fine subdivision home in those days for the price of one of today's twenty-foot, two-hundred-fifty horse, metal-flake, carpeted, aquatic missiles, outfitted with enough electronic geegaws to operate the space shuttle. But even by the comparatively innocent marine standards of 1972, Ray's boat was bare-bones basic, as was mine back at home. Certainly adequate for the job, but hardly the type of craft you'd expect to see on the cover of *Outdoor Life* magazine.

"I use this fancy boat when I've got guests," Ray said matter-of-factly. "I'd really rather fish out of my little boat, but I don't think you're ready for it yet."

"What little boat?"

"Forget I mentioned it," he replied. "You don't need to learn too much at one time."

We clamored aboard and I set about the task of organizing my inventory. With any luck, I'd be ready by the time we motored to our first spot.

Not quite.

Unbeknown to me, we were already at the first spot. Or nearly so. Ray goosed the electric motor toward a downed tree near the launch ramp, then turned it off just as quickly.

Although they are called "trolling" motors, these battery-powered workhorses are rarely used for that purpose. Instead, they are mounted on the bow of a bass boat and switched on and off to facilitate pinpoint positioning of the boat. It's a task much more easily accomplished from the front of the boat, not the back. Yet here was Ray, working it like an outboard.

I asked why.

"'Cause you need to be up there in pole position," he answered nonchalantly, digging through his cigar box.

Pole position?

What an odd use of the term. Was this a NASCAR event or bass fishing? And yet the malapropism sounded perfectly normal coming from Ray's mouth. It actually made sense.

"What sort of top-water plug you gonna use?" I inquired.

"Oh, just a little somethin' I put together," he answered.

With that, he produced the strangest amalgam of feathers, plastic, spinner blades, and treble hooks I'd ever seen and attached it to the end of his line with a simple open-and-close metal snap.

"Doesn't it have a name?"

"A bunch of 'em, I reckon. I got all these parts off of some old baits I had at the house."

Ray cocked his arm and flicked his wrist. The lure shot off the end of his rod in a trajectory as flat as a high-powered rifle. It touched down approximately three feet beyond the last exposed tip of the tree, just where the branches submerged. The instant it landed, he began cranking the reel handle rapidly.

What happened next was better heard than seen.

The thing came churning back—bubbling, gurgling, sputtering, and chopping—like an egg beater on steroids.

The evolution of fishing techniques takes time. Its various stages go by a variety of names. The technique Ray was demonstrating is known today as "buzzbaiting." You can walk into any sporting goods store in Dixie and find entire racks of buzzbaits produced by any number of manufacturers. Certainly a few vintage casting lures were capable of making such a splashy racket back then, but mostly they were homemade. In the hands of practiced veterans like Ray, they had been producing bass for decades. But if you walked up to any retail counter in the early 1970s and asked for a "buzzbait," you would have been greeted with two words:

"Say what?"

I watched in fascination as the lure came skittering back to his rod. Ray flicked his wrist again and sent it back to the same spot. The same *exact* spot. I'll bet it didn't hit a sixteenth of an inch off the earlier mark. Back it came, leaving a bubbly wake like a frightened hatchling wood duck.

Again he cast.

Again he retrieved.

This went on ten, twelve, maybe fifteen times.

"I don't believe anybody's home," I suggested, a reference that could, if one so chose, apply both to largemouth bass in submerged trees and gray cells in a certain cranium in the rear of the boat.

"Oh, he's there, all right," said Ray. "He just ain't interested yet."

Ray twisted the knob on the electric motor for two or three seconds, just enough to push the boat slightly forward and to the left. Then he turned it off and cast again, retrieving his lure from an ever-so-slightly different angle.

Still no strike.

He cast again.

And again.

And again.

"You really believe in wearin' it out, don't you?" I said.

"Oh, I've just begun," he replied. "See, I been a'comin' through the livin' room with that bait all this time. I reckon that ol' bass ain't in the livin' room right now. He may be in the kitchen, rumblin' around for breakfast."

Ray nudged the electric motor once more for a new position. We are not talking a vast change. Perhaps four or five feet and a half-dozen degrees of angle. Then he went back to frothing the surface with that ungodly concoction of feathers, plastic, and steel. I rigged a weedless plastic worm and began plumbing the bottom.

"He's not out there in deep water," Ray offered. "He's somewhere in this tree."

"I dunno," I responded with the foolish confidence of a sophomore. "Everything I ever heard or read says bass are deep this time of year."

"*Herr-herr-herr-herr!*" Ray laughed. "You been listenin' to the wrong people and readin' the wrong stories. Sounds like the stuff you sportswriters are always tellin'."

He continued to cast. And retrieve.

And cast. And retrieve.

It was getting ridiculous. I would have motored away ten minutes ago.

"Wel'sir, he ain't in the kitchen, either," Ray finally announced. "He may be like you."

"Wha'daya mean?"

"Why, this ol' bass may be another late Sam Venable," he said. "I bet he's still in the bedroom, sound asleep. D'you always oversleep like you did today?"

Ray didn't wait for a protested reply. "His bedroom is a little harder to reach than the kitchen and livin' room," he began to instruct, adjusting the boat's position again. "You see how that limb hangs down over yonder? Well, I figure that's where his bedroom must be."

"How do you expect to cast into there without hanging up?" I wanted to know.

"Like this," he said, unleashing another dart.

If I hadn't seen what happened next with my very own eyes, I wouldn't have believed it. In fact, I'm still not certain I believe it, even though I had a fifty-yard-line seat to watch the action.

The track Ray's lure was on would have carried it directly across the overhanging limb and into a jungle of vegetation. But just before it plummeted into disaster, he flicked the tip of his rod and dropped the line—the ski rope, that is—roughly half an inch. As if directed by remote control, the lure faltered in flight like a butterfly, shot underneath the limb, and touched down right on cue. It was a masterpiece of accuracy, if not sleight-of-hand.

Ray began turning the crank on his reel. As the lure gurgled along through the tiny opening, he chanted, "Come, butter, come."

He grinned sheepishly, keeping his eyes transfixed on the gurgling lure.

"You gotta coach a fish along sometimes, same as makin' butter."

The coaching apparently was not appreciated down below. Neither that offering, nor the three that followed, produced results. But watching Ray's marksmanship demonstration with every cast was worth the price of admission.

On the fifth retrieve through the tiny aperture, however, a bass reacted as if Vince Lombardi had just provided the coaching stimulus. There was a swirl. And then a hole exploded on the surface.

"I *knew* he was still in bed!" Ray cackled with glee, setting the hook and leaning into the rod. "He just needed wakin' up! *Herr-herr-herr-herr!*"

The largemouth wallowed momentarily on the surface, then attempted to tunnel into the bowels of the tree. No doubt it had escaped this way before. No doubt it had never been attached to ski rope.

"Don't just sit there!" Ray hollered to break my trance. "Get the net!"

The bass leaped skyward as I dove toward the floor of the boat, grasping frantically for the net's tubular handle. I managed to shove the mesh beneath the surface just as Ray led the fish toward the side of the boat. One scoop and it was ours.

It would not be wholly correct to describe this bout in the traditional sporting sense of "playing the fish." "Well-roping" was more like it. The whole thing was over in an eye blink. The bass was likely more confused than whipped.

We both guessed it at three and a half pounds. Maybe four, given the excitement of the moment. I began wrestling the hook loose while Ray reached for his "livewell," a rope stringer. As the fish was being lowered overboard—on a cord even more stout than his monofilament ski rope, by the way—I inspected the lure closely.

The body had come off an old Arbogast Hula Dancer plug. The blade, best I could tell, was robbed from a Go Devil spinning lure. The needle-sharp treble hook was dressed in yellow and red feathers of unknown origin. Along the wire shaft, just below the blade, Ray had attached a one-eighth-ounce bell sinker to give it more weight for casting.

"You really *did* make this thing yourself, didn't you?" I said.

"Of course," Ray answered with a shrug. "You don't think I could buy anything like that, do you?"

"Reckon there might be another one in that box of yours?"

My question evoked yet another horse whinny.

"Normally not," he quipped. "As you can see, I travel pretty light. But let me poke around. You never know what I might find."

Ray held the box close to his face and raised the lid slightly. He peeked inside cautiously, like maybe it held a captive bird. Slowly, he rummaged through the contents with one of those sausage fingers.

"I don't know why I'm doin' this," he said dryly. "You'd think a sportswriter with all of your connections would come out here equipped to fish. But I reckon you're my charity case today. Here."

With that, he tossed one of the sputter lures toward my end of the boat. It hit on the seat with a clang. I picked it up and tied it on my ski rope and, later in the morning, finally caught a largemouth bass on it. My fish was slightly smaller than the one Ray had just seduced, and it came after hundreds of casts, most of them errant. My proficiency at snagging brush and looping the lure over limbs was fast becoming legendary.

I still own that lure, plus a couple of others I managed to beg off of Ray that morning. I can cast them with far greater accuracy these days. But they never leave my tackle box, despite the fact they could attract bass just as readily today as they did thirty years ago. If I want to throw a buzzbait for bass, I use a commercial model. Commercial buzzbaits are replaceable.

Only a fool would risk losing a national treasure.

"Rock-
Elephant"

Since the statute of limitations has surely passed, I'm going to admit a breach of occupational protocol: I conducted far more "field research" than absolutely necessary to produce the first story I ever wrote about Ray Hubbard's bass-fishing prowess.

If pressed about the matter back in 1972, I surely could have baffled my boss with all manner of obfuscation about the necessity of revisiting the source of notes, mental and physical, I had made on our initial trip together. I could have insisted my first photographs of Ray in action had not met my personal and professional standards and that a return trip was required to take advantage of better lighting. I could have said his approach to fishing was so unusual and so effective, I needed proof that the first venture was not a fluke.

And it would have been a grand lie.

The truth of the matter is, our initial trip produced all the notes and photos I needed to crank out a good story. The reason I telephoned Ray and made another appointment was because I had enjoyed his company and wanted another demonstration of his top-water genius. My position with the newspaper afforded me many hours afield with some of the best hunters and fishermen in eastern Tennessee, and I enjoyed the experiences immensely. Yet nearly all of these outings were—don't laugh; I'm serious—vocational assignments. After the story was finished, I'd start looking for new horizons.

But this character was different. There was something about his friendly smile, his country boy demeanor. Not to mention those sudden outbursts of *"Herr-herr-herr-herr!"* He should have had the sound patented. He was absorbed in his passion for bass, yet totally void of pretense. What you saw was what you got. By the time my first account of our adventures was published—Sunday, July 30, 1972, page D-7, in the *Knoxville News-Sentinel*—Ray and I had already begun to form a friendship.

Sociologists call this "male bonding," which is the most ridiculous-sounding term I've ever heard—unless the men in question have a collective interest in concrete. When I hear someone discuss "male bonding," I conjure mental images of Alan Alda clones sitting bare-chested around a campfire in some suburban backyard, beating tom-toms and complaining about how boyhood was an incomplete experience for them.

Horse manure. Ray and I simply were two bass anglers whose personalities went together like chicken and rice.

No doubt a very practical matter was our weird off-day schedules. Whenever I hunted or fished on a busman's holiday, I often did it alone because most everyone else I knew was sitting behind a desk on Tuesday. And then along came a fellow whose free-time schedule meshed perfectly with mine. Like the old TV beer commercial used to say, it just doesn't get any better than this.

Ah, yes. Beer.

Ray and I had been fishing together for maybe a month or two when we decided to spend a Tuesday casting the coves and flats of the George's Creek embayment of Fort Loudoun, near his home in Louisville, Tennessee. Don't pack a lunch, he had told me. The tomatoes were about to take over his garden. He was going to bring all the fixin's for sandwiches—meat, bread, condiments, the works. We'd eat till we busted.

"Well, let me bring *some*thing," I told him over the telephone.

"All right," he said, "you bring the drinks."

Before leaving the house early the next morning—not early enough for Ray, naturally—I stuck four or five soft drinks into a cooler full of ice. Just as the refrigerator door was closing, however, I spotted the remnants of a six-pack of beer.

"Those oughta go good with sandwiches," I said to myself. And in they went, too.

It never dawned on me that I was about to bring pork to the synagogue.

The subject had simply never come up. Up until that point, approximately 99.99 percent of the conversations that Ray and I shared on the water revolved around large-mouth bass. We really hadn't gotten to know each other all that well. Other than the fact that he fished on Tuesdays and sold and serviced sewing machines on Mondays, Wednesdays, Thursdays, Fridays, and Saturdays, I didn't have a clue about his private life. Nor he mine. Neither of us had discussed where we spent our Sundays. Like beer, the subject had never come up.

It should have dawned on me that something divine was at work here, for from day one, Ray would toss an occasional Bible verse into his stories. Indeed, one particular sermonette was delivered with regularity. It always came when I would grow frustrated with his incessant casting—over and over and over and over—into the same spot, without a strike.

"You ain't gonna catch a bass there, Ray!" I'd finally shout. "How many times are you gonna do that?"

He would just laugh.

"Now lissen—in the eighteenth chapter of Matthew, you'll find the story about the time Peter asked Jesus how many times he was supposed to forgive his enemies. Was it supposed to be seven times? No, sir. Jesus told Peter to forgive his enemies seventy times seven! And that's what I'm doin' with this bass. I've got t'forgive him seventy times seven for not strikin'. *Herr-herr-herr-herr!*"

And back in his lure would go.

Another Hubbardism I immediately noticed was Ray's even temperament, especially when times got testy. As delightful and relaxing a pastime as it can be, fishing offers a variety of opportunities for intense frustration—everything from backlashed reels that look like teased hair from the 1960s, to underwater structure that snags a prize lure and stubbornly refuses to let go, to the ubiquitous big one that inevitably gets away. Ray experienced them all from time to time. But he never let them interfere with his fun. A pained expression and a moan were about as vile as he would get.

I, on the other hand, have been known to turn the air blue.

It's either a flaw in my character or the result of spending more than thirty years in newsrooms, particularly in the mass confusion of Type-A personalities at deadline. Some of the most awful cussings in the history of the spoken word used to be uttered then, by women as well as men. Newsrooms have grown decidedly quieter since desktop computers replaced the clackety-clack of manual typewriters, but be not fooled. If you are easily offended, stay away from any newsroom when an entire computer network goes on the fritz—which happens, oh, not more than once or twice a day in most operations.

Thus, suffice to say I barked more than an occasional "pshaw" and "ding-dang" when my fishing fortunes turned sour.

Ray didn't so much as blink. From day one, he had a constant comeback to any of my explosions: "If you want to say something hard, say 'rock.' If you want to say something big, say 'elephant.' Don't go on usin' those other words. All they do is make your blood pressure rise. Just say 'rock-elephant.' You'll feel a lot better."

No. No! Hell, no! I did *not* feel better.

It was like trying to stifle a sneeze or silence a laugh in church. There are times—like when a knot you just spent two minutes tying comes undone—that a thorough cussing is not only warranted, it is expected. It got to be so predictable that when I would erupt like a volcano and Ray would issue his "rock-elephant" advisory, we'd both wind up laughing at our impasse.

So there we were that day, in the back of George's Creek, in the heat of mid-day, when Ray motored over to shady bank, dropped anchor, and announced lunch was about to be served. He set his cooler and a brown grocery bag atop the plank seat and began to spread the table.

Out came a loaf of bread. Then a pack of lunchmeat. Then a jar of mayo, along with packets of salt and pepper, and three of the plumpest ripe tomatoes this side of a produce commercial. Two were ruby red, the other one yellow.

Digging deeper into the grocery bag, Ray brought forth a kitchen paring knife, wrapped for safety in several sheets of paper towels. With it, he peeled and sliced one of the red tomatoes. Satisfied all the ingredients were in place, he began assembling gargantuan sandwiches.

He handed the first to me. It must have weighed nearly one pound. I started salivating like one of Pavlov's dogs as I opened my cooler and ripped the pop-top off a can of beer.

Even though the sun shone brilliantly, the mood immediately darkened. It may have been similar to the sense of doom Saul experienced on the road to Damascus, shortly before he was struck blind.

"*Beer!?!*" Ray thundered. "I sure don't want none of that! It don't even need to be in this boat! What in the world did you bring something like that along for?"

"To drink," I replied. "I figured you'd want one, too."

Suffice to say this was a miscalculation—dietary, social, and otherwise—of the highest order. For I was about to learn that Ray spent his Sunday mornings in a pulpit, and one of his pet topics was the evil of strong drink.

In some circumstances, this situation would not be the least bit awkward. One party would simply say "No, thank you," and that would be the end of that. Not in Southern fundamentalist religious circles, however. Here was a major faux pas.

"Gosh, Ray, I'm sorry to have offended you," I said. "I'll just pour it out."

"No, don't do that!" he snapped back. "That barley juice'll kill ever' fish in this cove! I know how that stuff'll rust nails. Why, there'll be fish floatin' up all over the place if you pour it in here. Besides, somebody'll smell it a mile away."

For once, I was the first one to initiate a belly laugh. I'd never heard the term "barley juice" before, certainly not the disparaging way he spit it out.

"It ain't quite *that* powerful, Ray," I said. "If it was, I could bring a six-pack down here and wipe out the entire lake!"

"And I reckon you would, too! Usin' that rye runoff is the only way you could catch a limit of fish!"

"Rye runoff" further tickled my funnybone. And, incredibly, the more I laughed at Ray's protests, the more amused he became with me. Even in the midst of his thunderations, that smile was starting to sneak back in.

"Well, if you won't let me pour it in the *lake*, I reckon I'll pour it in *me*," I said. "It's gonna spill all over the boat."

That one got the better of him. He was certain a gallon of pine oil cleaner wouldn't stand up to a twelve-ounce beer. And should a parishioner drop by unexpectedly later on and get a wiff of stale suds in the bottom of the preacher's boat— well, perish the thought!

"OK, but if you go t'actin' crazy, I'm takin' you to the bank!" he warned.

A shaky truce ensued for the remainder of the afternoon. Perhaps providence decided the best thing for each of us to do was get back to the business of bass fishing. Sermons about alcohol, from either end of the boat, ceased.

After a bit, I grew hungry again and offered to make us each another sandwich. Ray declined but told me to help myself to the plunder. As I was running the blade of his kitchen knife through the yellow tomato, I commented on how sharp it was.

"A knife ain't no good unless it's sharp," he said. "A dull knife'll cut you faster than a sharp one."

"True," I noted around bites of my sandwich, "but that thing's sharper'n any knife I ever held. It's like a razor blade."

Ray waited until I was chewing contentedly before he spoke again.

"Oh, I always keep that knife extra sharp. It's the one I use to cut my toenails."

I stopped chewing. The delicious sandwich inside my mouth suddenly had lost its flavor.

Did he say *cut his toenails?*

Ray milked the moment like a polished comedian. He just stared at me as the sandwich began to balloon in my jaws. Then he turned to make another cast.

"'Course, I reckon a man with a mouthful of barley juice don't need to worry," he sniffed. "All that alkey-hol will surely kill the germs."

He looked back in my direction and grinned broadly. I started chuckling. So did he. It was a collision of funny-bones. We both got so tickled I was afraid I'd choke. There was nothing to do but spit the sandwich overboard so I could draw a breath.

"Now just look at you, about to cause another fish kill!" he chided. "I'll swan, I don't know what I'm gonna do with you!"

I knew exactly what I was going to do. I was going to clear the next few Tuesdays for more "field research," even if it meant heaps of verbal abuse. A man could get used to this.

Operation
Code Name

Thomas here is a rule of Southern culture requiring nicknames among friends. Male friends, that is. Normally, this dictum does not transcend gender.

When Southern female friends get together, they address each other with the same name (usually a double-worded one) each was given at birth. You know you have crossed the Mason-Dixon line—and, more importantly, are pointed in the proper direction—when names like "Mary Ann," "Lois Evelyn," "Sue Ellen," "Barbara Jo," and "Betsy Louise" begin creeping into the conversation.

It is not mandatory for Southern males to be saddled with a double name—the addition of "Bob" to any moniker being the exception—because nomenclative formality is discarded approximately seventeen seconds after christening. Only agents of the Internal Revenue Service know Southern men as "Clarence," "Frederick," "Christopher," and "Randolph." Everyone, from the president of Rotary to the deputy sheriff, calls them "Peanut," "Noodle," "Dawg," and "Bones," respectively, and will use those names in every situation, formal or otherwise.

Which explains why Ray became "Herb."

Don't ask me where the name came from. It just seemed like a good handle for him after the newness of "Ray" wore off.

Nobody ever said nicknaming has to make sense. The guy down at the hardware store nicknamed "Tiny" usually tips the scales around three-fifty. "Curly," as any fan of the Three Stooges will attest, usually has a chrome dome. And "Porky" might be skinny enough to hide behind bean poles.

The name merely has to feel right. It either does or it doesn't. "Herb" was perfect in this particular instance.

Apparently I wasn't the first person to think so, either. After Ray's death, I discovered some of his brothers had called him Herb during their growing-up years. If his given name had been Herbert, they, and I, might have started calling him Ray. Go figure.

Thus, from the mid-1970s on, the only time I referred to him as "Ray" was in print. Every year or so during my tenure as outdoor editor, I would crank out a feature story or column about one of our adventures. This habit continued after 1985, when I left the outdoor beat and began the humor column I still write four days a week. There was also enough natural-born ham in the guy that he became my model of choice when I needed a subject for a photo spread. Over the years, pictures of Ray Hubbard in action illustrated many magazine features, including ones for *Bassmaster, Tennessee Conservationist, Tennessee Sportsman,* and *Southern Outdoors.*

But anytime I wasn't typing his name, he was "Herb" or "Herbie."

I became "Bill D." and "Fingernail." Unlike the mysterious "Herb," at least there was an explanation behind the nicknames he gave me.

"Bill D.," of course, was a reference to legendary professional bass angler Bill Dance. Not that I was even *remotely* in Dance's league as a fisherman, you understand. Instead, Ray— er, Herb—was watching Dance on television one time and noticed that he approached a fishing site a bit faster than Herb thought necessary. From that point on, anytime I was at the controls of the motor, I answered to Bill D.

Herb believed in stealth while he was fishing. He wore dark clothing, sometimes even a camouflage shirt, when casting over the shallows. Nothing would irritate him more than when I'd show up wearing a white T-shirt. And when it came to approaching a likely looking spot, anything beyond no-wake speed for the last hundred yards was heresy.

"You're goin' too fast, Bill D.!" he would advise. "Slow down! Slow down!"

This critique was issued faithfully, even if the only force pushing the boat was wind. I do believe if we had launched at

Fort Loudoun Dam, with intent to fish twenty miles upstream, he would have started yelling, "Slow down! Slow down!" the moment we cleared the no-wake zone at the ramp. If nothing else, Herbie was predictable in his admonitions.

"Fingernail" came much later, after Herb was starting to slow down from heart disease, kidney failure, and some of the other ailments that progressively weakened his body.

"I ain't never figured it out," he groused to me one day. "Here the Good Lord has given me all this pain and sufferin' and misery, and he wouldn't harm a fingernail on your hand."

Thus, Fingernail it was to the end.

But humanized nicknames were only a part of Herb's vocabulary. He was a master of codes, firmly believing "something" would jinx an outing unless all details were camouflaged. He should have worked for the CIA or FBI. J. Edgar Hoover would have loved him, for Herb absolutely, positively knew there were evil forces out there determined to riddle our every plan with fistfuls of monkey wrenches.

It was for that reason that he and I quit "going fishing." Instead we "played golf."

"Ever'time you call me on the telephone to go fishin', it winds up rainin'," he grumbled one day. "It's got to be your fault. This never happened before I started fishin' with you. So forget that word 'fishin'.' We ain't ever 'goin' fishin'' again. We're goin' to 'play golf.' Got it?"

I did as directed on our next journey. Every word of our conversation revolved around carts and clubs and little white balls. Just as we pulled up to the launch ramp—I mean, first tee—the heavens opened. Rain came down in sheets.

Herb turned toward me, glared like a school teacher who'd just been spit-balled from across study hall, and barked, "All right! Who'd you talk to? I know you had to tell somebody we were goin' fishin', or else this wouldn't have happened!"

It finally got to the point that our destinations became coded.

Norris Lake, he declared one day, would hereafter be "Sirron." That's Norris spelled backwards, if you haven't guessed. Herb detested fishing the clear waters of Norris and would only join me there in the fall, when spotted bass were slashing recklessly into surface lures. Anybody who'd spend much time on that lake has got be backwards, he decreed.

Douglas Lake became "MacArthur" because we "generally" caught bass there. (MacArthur was an army general, get it?)

After hearing stories of a nine-pound bass from Cove Lake—and immediately dismissing them as fabrication—Herb evermore called it "Liar's Cove."

One of the small farm ponds we regularly visited didn't even have a name, formal or otherwise, until Herb christened it "PTL"—Pool Table Lake—because of its diminutive size.

Another small lake actually had a name. But because he favored the use of acronyms whenever possible, Bluegrass Lake forever became "BGL."

Fort Loudoun, beset for decades by pollution, was "Fort Nasty" or the "Cess Pool." Watts Bar reservoir was simply the "Bar," and you can bet your bottom dollar it was the only bar he ever visited.

Herb even began code-naming various sections of individual lakes and rivers. The Muddy Creek embayment on Douglas—make that MacArthur—became "Watermelon Hollow" after he and I split, and ate, an entire watermelon there one afternoon. A launch ramp on Fort Loudoun, where a couple of happy chaps had once offered to help unload the boat, was dubbed "Friendly Hollow." Another section of the lake, where we had dined on delicious wieners from a nearby country store, turned into "Hot Dog Hollow." And on Little River, there was "Boots Island," a name not recorded on any map, but the very place where he and I found a pair of old-time brogans atop a rock.

Once a place was christened in "Herbese," the name became permanent. The junior member of the tandem either had to learn it quickly or become hopelessly lost. If the command, "Watermelon Hollow, noon" was on my phone recorder, I had dang-sure better know where to be—even though, according to Herb's watch, I was always late.

Sometimes the codes weren't intentional because Herb never met a word he couldn't butcher. Then again, maybe he was simply trying to improve upon the King's English. Whatever the reason, I became so fluent in Herbese that translations rarely were required. The day he mentioned the "emporium" in Chattanooga, for instance, I knew immediately he was talking about the city's impressive new aquarium.

During my years of nicotine addiction, I never chewed "tobacco" or smoked "cigarettes." At least not to hear Herb tell it. They were "black leaf forty" and "cancer sticks," respectively. As any tobacco chewer will attest, bits of stem occasionally show up among the leaf, necessitating a quick *"pp-thoo"* removal. Herb never missed the opportunity to sermonize on the economics of the matter: "Why any man would spend good money just so he could spit crossties out of his mouth is a mystery to me. Don't you know the tobacco companies buy used crossties from the railroad and grind 'em up and mix 'em in with that black leaf forty? And you're silly enough to fall for it! Beats all I ever saw."

After one of his many hospital visits, Herb announced his breathing had been so labored in the emergency room, he required "octagon." And once when he nearly choked on a bit of food, he told me technicians had revived him with the "hymen maneuver." I understood completely.

Down-home meteorology was yet another subject regularly treated to his special words, phrases, and sayings. "I don't know why you complain so much about the wind a'blowin'," he would constantly remind me. "What would the wind do if it didn't blow? That's why they call it *wind.*"

Herb was positively convinced storm fronts out of the west often missed Tennessee completely because they were split apart—one fragment going north, the other south—by the "plow point" on the northwest corner of the state. "Just look at a picture of Tennessee on any map," he insisted. "That plow point's right there in plain view."

But one day he threw a double curve at me. It occurred as we were navigating rural Chapman Highway, en route to MacArthur. Originally a two-lane rural affair, Chapman Highway had been widened to four lanes half a century earlier—back when ten cars every thirty minutes, traveling at the blinding speed of forty miles per hour, constituted rush hour. As we weaved in and out of the zip-zip traffic, I happened to comment on how dangerous the situation was.

"I wish they'd do something about this highway," I complained. "If it only had a median, that would help. I hate to see those other cars comin' at me so fast with nothin' but air between my lane and theirs."

"Puttin' in a medium would take too much time and cost too much money," he countered.

"Median," I corrected.

"That's what I said—medium," he fired back. "Ain't you listenin'? Anyhow, all the state needs to do is hire an extortionist to take care of the problem."

"What's an extortionist got to do with it?"

"Somebody to cast out the devil. This highway is cursed. That's why there are so many wrecks on it."

"By any chance, do you mean an *exorcist?*" I asked.

"Right. An extortionist. Did you see that movie?"

Hoo-boy.

Food ranked just slightly behind fishing in importance to Herb, and he knew how to pack it away with techniques that might cause Julia Child to faint. In addition to frying the fillets of bass, crappies, and bluegills, he also chopped up pieces of the dear-departeds' tails and fried them as well: "Mmmm! Just like 'tater chips!" Then he would make "chocolate gravy"—yes, with chocolate milk—out of the skillet leavings. Aside from iced tea, one of his favorite liquid refreshments was the juice from pork and beans, guzzled straight-up through small holes he had punched in the top of the can.

But since he also had a taste for commercially prepared meals, Herb began assigning code names to the chow houses we would visit. Ramsey's Cafeteria was the "Feed Trough" because we ate there so often. A nearby Ramada Inn that featured cherry cobbler among its buffet selections was officially recorded as the "Cherry Pit." He loved to stop for breakfast at any "Awful House" franchise. Conversely, he wasn't overly fond of roadside deli–service stations. These were "smokehouses," Herb claimed, the food inside invariably tainted by an ever-present blue haze from those infernal "cancer sticks."

Yet my all-time favorite code name was the one he laid on the Wagon Wheel Restaurant in Dandridge, Tennessee.

Herb and I happened to be on MacArthur the afternoon of October 3, 1995, when a jury in Los Angeles was to return its verdict in the O. J. Simpson murder case. I intended to write a column about the verdict and told Herb I positively *had* to be sitting in front of a television set at the appointed hour. We came off the lake, drove to Dandridge, and joined what appeared to be half of Jefferson County in front of the eatery's diminutive black-and-white TV.

The Wagon Wheel isn't a large establishment in the first place, and I'll guarantee it had surpassed the fire marshal's capacity by forty heads. We barely were able to squeeze into the front door. Finding a table was impossible. The TV was perhaps thirty feet away, atop the luncheon bar.

"I gotta get closer," I told him. "I can't miss a second of this."

With that, I hit the floor and crawled, literally, between legs and feet. Periodically, I would surface in the sea of humanity—like a groundhog sitting up in a bed of tall clover —until I could get my bearings. Then back down I'd go.

The seconds were ticking down when I finally surfaced near the television. Judge Lance Ito's clerk was about to read the pronouncement to an international audience of multiplied millions, not to mention the throng at the Wagon Wheel.

A hush fell over the restaurant crowd. But even in the near-silence, the tiny TV was barely audible.

"Turn it up!" someone yelled from the rear of the room.

As I was nearest the set by now, I reached up, grasped a knob, and started to rotate it.

At least a dozen voices shouted in unison, *"NOT THAT ONE!"*

I looked up in horror and realized that instead of the volume control knob, I was about to change channels. I jerked my hand away from the set and cowered on the floor. It's ten thousand wonders I wasn't summarily ejected out the side window.

The now-famous verdict of "not guilty" was announced amid jeers from the audience. I scribbled hasty commentary from some of the assembled masses as they began to drift out of the joint. Herb, who had kept his distance throughout, sauntered up and wisely suggested we might want to grab a bite elsewhere, given my precipitous decline in the popularity poll. I concurred.

In the parking lot outside, he unbraided me. "You crazy idiot! If you'da changed that channel, they'da mopped up the floor with you!"

"Wouldn't my ol' buddy have come to my rescue?" I asked, feigning childlike innocence.

"No, sir!" he said. "You'da deserved it."

Then he smiled. "But I woudda stayed back there and prayed *real* hard for you while they were stompin' your guts out. *Herr-herr-herr-herr!*"

From that day forward, any time we were within twenty miles of Dandridge near mealtime and Herb suggested we dine at the "Floor Crawler Cafe," I knew exactly where we were headed.

Up Close
and Personal

Herb was on the telephone, and he was on a mission.

"The problem with you, and about 99 percent of other people who call themselves bass fishermen, is that you spend most of your time castin' for the end of the rainbow when it's right there under your nose. You're just too blind to see it."

I wasn't exactly certain where this lesson was headed but knew better than to inquire. All would be revealed in due time. When Herb climbed into his piscatorial pulpit, you might as well settle in and make yourself comfortable. The sermon was going to wind on for awhile, but it was always worth listening to. And when it came time for the altar call—a gathering at the river, as it were—you'd better have one hand on a rod and reel and the other on the steering wheel. The preacher didn't like to be kept waiting.

"Right before I came off the water the other afternoon, I was watchin' two fellows in a big bass boat make fools out of themselves."

Now, wait a minute! That comment demanded elaboration, and I was just dumb enough to ask for it.

"I didn't know you were goin' fishing," I protested. "You didn't say nothin' to me about it."

"Sometimes it's better if I go alone and experiment," he answered immediately. "I'd hate to waste your valuable time on a new place."

"So I guess I'm about to hear another one of your ancient history lessons, right? One of those you-shoulda-been-there-yesterday stories?"

"Why, I can't imagine you'd say such a cruel thing," Herb continued, his voice dripping with sarcastic delight. "You know if I took you somewhere and it turned out to be a dud, I could never forgive myself."

I cut right to the chase: "So how big was your best one?"

"Oh, it was small," he sniffed. "About five pounds. I'm sure there's a better one in there waitin' for you—but if you don't quit interruptin', it's gonna be winter, and the ice will be four inches thick, and we'll have to wait till next summer to try him out. Now, where was I?"

"Somethin' about two guys in a bass boat."

"Oh, yeah. They put on the best demonstration of how *not* to fish I've seen in awhile. I was putterin' over to this tree I knew about, but before I could get there, they come a'runnin' in, full bore. Throwed water ever'where. Sorta reminded me of you. Anyhow, they went t'flailin' away with spinnerbaits. But the trouble is, they didn't ever cast where the fish oughta be. It was like they were afraid of hangin' up, and anybody knows you oughta not ever hang up a spinnerbait.

"Wel'sir, I moseyed over to an old boat dock that was pretty close by and fished there awhile. But I watched 'em out of the corner of my eye. I bet they didn't stay at that tree more'n five minutes. Then—*whoom!*—they fired up that big motor and off they went in a big rooster tail. By the way, that wasn't you, was it?"

"No," I said dryly. "I haven't been fishing in over two weeks. I thought my dearest friend in all the world was gonna call and ask me to go, but I guess I was wrong."

"Oh, I called two or three times, but your number was always busy. So I just figured you were too tied up with work."

It was useless. Going one-on-one with Herb was like trying to argue with the captain of an international championship debating team. The harder you worked to pin him down, the easier he escaped. He could volley two totally different replies to your question before the original query formed in your mouth.

Take the time Herb became obsessed with the notion of owning a barn to store his angling junk. For months, whether we were on the road or on the water, whenever we'd pass by a stately old barn, he would say, "Oh, I'd love to have that thing. Ever'body's got a barn but me. Why can't I ever have an old barn?"

Finally, I said to him, "Sounds like to me you're coveting another man's possessions. Doesn't the Bible warn against that?"

Herb, as usual, produced an evangelical ace from deep within his sleeve.

"I am *not* covetin' another man's barn!" he shot back. "You see, I want him to have a *better* barn than that. Then I'll just take the old one off his hands so he won't have to mess with it."

That's what I mean about the futility of trying to trip him up. It was a waste of time.

"So what'd you do after the guys in the bass boat finally left?" I asked.

"First, I let the water settle down for a little bit," he said. "Then I just eased over and ran a Jitterbug right along the side of that tree. That ol' bass was laid up there, just like I figured he'd be. *Herr-herr-herr-herr!* It was bettern' gettin' money from home without writin' for it. You wanna hear from that bass right now?"

"How can a bass talk to me?"

"Just lissen real close."

I could hear the receiver moving through dead space. Then it was filled, ever louder, with a crackling, popping noise. It didn't take but a second to recognize it: the sound of fish frying in a skillet.

Herb returned the telephone to his ear.

"Don't that sound delicious? Now, if you'uz over here, we could do some real eatin'. But you're always too busy workin' to eat, so I'll just have to enjoy this bass all by myself."

He hardly stopped for air.

"It's just as well you didn't eat, though. You'd probably founder on all this food, and then you wouldn't be worth a flip tomorrow. You need a lesson in how to fish for bass in close cover. Real close cover. A lot closer than you and me have ever fished together before. Meet me at the Golf Course tomorrow at noon. Surely you'll be awake by then. Besides, it needs to be good and hot where we're goin'. These bass don't worry about a little heat like some people I know."

"Do I need to bring anything special?"

"You know somethin? You're gettin' smarter! Now that you asked, yes, there is somethin' you can bring. You got a small saw?"

"What for?"

"That's not what I asked," Herb retorted. "I asked if you had a small saw."

"Yeah, I got one. It even folds up. I use it in the woods when I'm building deer stands."

"Oughta be perfect," said Herb. "Please, please try not to be late."

I arrived at the Golf Course at 11:55 on the dot. Herb, naturally, was pacing and looking at his watch.

"I'm gonna start tellin' you ten o'clock when I mean noon," he snapped. "Maybe that way, you'll be on time."

That tongue-lashing was no surprise. What was surprising, however, was the boat waiting at the end of the launch ramp.

Truthfully, "boat" isn't all that descriptive. "Fiberglass bath tub" is more like it. Indeed, that immediately became my code name for it—the Hub Tub.

You've seen those molded, two-man fish-and-hunt mini boats? This was its smaller cousin. Bow to stern, it measured eight feet. There was an electric motor mounted on the transom. A single shelving board amidships was supposed to hold both of us. In the floor lay an old wooden bolo paddle with "Mrs. Reed" printed in faded ink letters.

"This is the little boat I been tellin' you about," Herb announced. "I normally use this thing all by myself. But if you'll sit still and not turn us over, maybe—*maybe*, I said—we can manage to catch a few bass."

Fearing the worst, I gingerly boarded and sat down. One leg in, one leg out, Herb cautiously pushed us off the concrete. Then he made a lunge for the seat.

"Move over!" he hollered. "And quit rockin' the boat! Do you have to jerk and kick all the time?"

In point of fact, I was not jerking. Nor kicking. I was not even blinking. Indeed, I barely was breathing. If I could have inhaled through the pores of my skin and circumvented lungs and nose altogether, so much the better. Imagine sitting astride a Popsicle stick with someone else.

And we—gulp!—were actually going to move about? And cast?

Happily, the task was not nearly as difficult as it seemed. Within a hundred or so yards from the ramp, Herb and I managed to gather our sea legs. Or sea seats, as the case may have been. We stopped at one rocky point, made a half dozen casts, and convinced ourselves that capsizing—although not an impossibility—was certainly not as likely as it seemed only a few minutes earlier.

Satisfied we weren't candidates for the next day's obituary pages, Herb pointed the Hub Tub toward the mouth of a narrow cove that snaked off through the bushes. As we hummed along under the power of the electric motor, Herb relayed the small boat's history.

Seems it had been manufactured as the life raft for a fifty-foot houseboat. Herb purchased it out of a mail-order catalog twelve years earlier and had been prowling shallow bays and coves with it ever since. Early on, he tried powering it with a two-horse outboard, but that proved to be excessive. The electric motor was more its style.

This definitely was not the sort of boat to be taking into the wide, open spaces of a major reservoir. It wasn't built for speed, certainly. Nor for comfort. But for poking in and around the tightest of bass haunts, this was a jewel.

"Besides," he noted, "where else do you know of a boat that'll also serve as an umbrella?"

"Huh? Umbrella?"

"Yep. I got caught out in a bad storm once. No way I could get back to the car. So I just pulled my yacht up on the bank, turned it over, and sat underneath it till the storm passed. Then I went back to fishin'."

"Who was, or is, Mrs. Reed?" I asked, pointing to the paddle.

"She musta been a school teacher," Herb replied. "I reckon this was her paddle for whuppin' students. I got it for a quarter over at the flea market. Makes a great paddle if my motor quits workin'."

From that moment forward, any paddle in any boat on any body of water was not a "paddle." It was "Mrs. Reed." In the summer of 1994, when I was in my driveway going over all the accessories of an eighteen-foot bass boat I had just purchased from a dealer, I discovered a shiny new wooden paddle mounted in a storage locker. I immediately hopped out of the boat, marched into the garage, found a felt-tipped marker with indelible ink, and inscribed the words "Mrs. Reed" across the face of the paddle. Only then was the ship worthy of being launched on its maiden voyage.

When I say the cove Herb and I fished that first afternoon in the Hub Tub was "narrow," I mean it in the most literal definition of the word. He and I could work either bank at will. We didn't so much "cast" our lures as we did "toss" them. I doubt more than ten yards of line snaked off my reel the entire afternoon. Because we were operating in such a confined space, it was virtually impossible not to swim our plastic worms, buzzbaits and spinnerbaits over, under, around, and through every square inch of cover.

We didn't catch that five-pounder's brother, as Herb had hinted we might. Not even out of the same brush pile where Mister Big had originated. We caught a few bass, all right. In the two hours it took us to comb the first half-mile of the winding cove, we managed to put five large-mouths—one- to two-pounders—on Herb's stringer. But then we came to the end of the cove.

Or so it seemed.

"You got that saw?" he asked.

"Yeah. It's under my feet."

"Well, climb out when I shove us under this tree and cut a few limbs."

"How come?" I wanted to know. "Doesn't this slough end here?"

"Oh, goodness no!" he replied. "Goes for another three or four hundred yards. I been fishin' in here for years. But here's as far as I got the other day 'cause this sickeymore tree had blown down over it."

Herb gunned the electric motor—a nautical oxymoron, now that I think about it—and pushed the Hub Tub's bow into a wall of sycamore limbs and leaves.

"Now, get out, *and be careful!*" he coached. "The last thing I need you to do is turn us over back in here where nobody'll find us till the crawfish clean our bones."

The tree leaned over the water at about a forty-five degree angle. I'm still not certain how I executed the maneuver, but somehow I managed to pull myself onto the trunk and establish a semblance of a perch.

Herb backed off and pointed at one large limb with the tip of his rod.

"Get this one," he instructed.

In fifteen or twenty strokes, the metallic teeth did their work. The limb fell into the water with a quiet splash. Herb dragged it aside and tossed it onto the bank. He repositioned the Hub Tub and pointed again.

"Now, this one."

So done.

The scene was repeated. Again. And again. It took the longest, most arduous twenty minutes of intense labor in my life, but I finally carved a hole large enough for the Hub Tub to slip through effortlessly.

"Wait up there just a second till I get the boat turned around and my line straightened out," Herb spoke over his shoulder.

What else was I going to do? My monkey skills had gotten a decent workout in the previous twenty minutes, but I wasn't up to the task of following Herb down the slough, limb-to-limb.

Then I heard him cast.

Sweating, tottering, and cursing, I managed to turn myself around on the sycamore trunk just in time to see him drag a buzzbait across the submerged roots of a stump up ahead. There was an explosion on the surface. A big one. No, it wasn't that five-pounder's brother. But it was a good three pounds, maybe three-and-a-half. And I had to endure every second of the battle from the tree.

"Damn you, Herb!" I hollered. "Come get me!"

"Shhhh!" he hissed back. "Don't get so upset. Just remember to say 'rock-elephant.' It'll make you feel better."

"I'm gonna kick the rock-elephant outta you if you don't get back here!"

Herb strung the bass. Then slowly, he backed up the Hub Tub until it was directly beneath me.

I slithered in and glared at him.

"Herr-herr-herr-herr! You didn't think I'd take you into uncharted waters without testing them first, did you?" he laughed. "Why, if you'da worked that hard and we didn't have nothin' to show for it, I never could have lived with myself!"

He looked me up and down.

"Boy, you sure are a mess. Can't you get into a boat without trackin' leaves and spider webs with you? Here I give you a fishin' lesson like you've never seen before, and what do I get in return? A messy boat, that's what. You beat all. *Herr-herr-herr-herr!*"

Yep. It was a lesson, for sure. A lesson I'll never forget.

And I'd pay a hundred dollars, cash money, to watch the ol' shyster teach it to me all over again.

Made from
Scratch

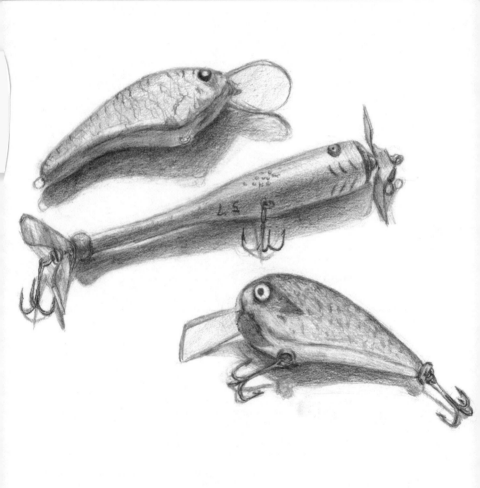

I t may not be the smoke-stacked manufacturing capital of the free world—thank goodness—but Tennessee is headquarters for a vast array of internationally known consumer goods. Everything from Saturn automobiles to Sea Ray boats, Goo-Goo Clusters to Moon Pies, Jack Daniel's whiskey to Wilson golf balls, all get their start in the Volunteer State.

There's another product that originated in these hills, one many people don't know about. Unless they fish for bass, that is. For members of this cult, the words "homemade plug" and "Tennessee" are synonymous.

In truth, not many are actually "homemade" anymore because hand-crafters can't keep up with the demands of the fishing public. Nonetheless, it is arguable that no artificial lure in the history of recreational fishing has had more of an impact—on the water and at the sales counter—than this family of chubby, balsa-wood crankbaits.

Exactly who started the craze will be debated as long as two or more opinionated anglers gather at the same place. Surely one of the first was a Maryville fisherman named Boots Anderson. He had a variety of names for his creations, including the "Top Secret" (for obvious reasons) and, later, the "Tennessee Shad." Throughout the 1960s and '70s, dozens of woodcarving bassoholic hillbillies—including Eddie Burke, Mike Estep, Dwayne Hickson, James Lovingood, Richard Perkins, Doc Daugherty, and B. B. Ezell—added their own twists.

Yet it was a carver from the "atomic city" of Oak Ridge whose fame spread across the waters of America faster than measles through first grade. Not only did Fred Young create an excellent lure, he also had the good fortune and marketing acuity to place it in the hands of some of the nation's top-ranked competitive bass anglers. Young also was blessed with a heavyweight brother named Odis, after whom he named his lure.

Once the "Big-O" hit the scene, bass fishing was never the same.

During a 1972 Bass Anglers Sportsman Society tournament on Tennessee's Watts Bar Lake, Big-Os were responsible for boatloads of fish. The lure's reputation grew exponentially, to the point that Young soon was months behind on orders. By 1973, tournament anglers were actually paying upwards of fifty dollars a day just to *rent* Big-Os from their owners—and woe be unto the unlucky caster who lost this precious commodity to a lunker or underwater snag!

That same year, a tackle manufacturer in Hot Springs, Arkansas, named Cotton Cordell bought the rights to the name Big-O and started production of the lure in plastic. Other "alphabet plugs" soon followed, including the Big-N from Norman Lures in Fort Smith, Arkansas; the Bee-O from Bumble Bee Lures in Athens, Alabama; and the Balsa-B and Big-B from Bagley Lures in Winter Haven, Florida. From the mid-1970s on, virtually any bass angler's tackle box has contained a model or two of these fish-catching marvels.

Herb owned dozens of original Fred Young Big-Os that never saw duty in the water. He had collected lures throughout his life, but when the Big-O hit the scene, he became obsessed with it and other "homemades" by Tennessee artisans. At the time of his death, he owned one of the most extensive collections of these plugs ever assembled. He also collected turn-of-the-century plugs from companies like Heddon and Creek Chub and vintage outdoor magazines like *Field & Stream* and *Sports Afield.*

But the collectibles were a mere fraction of the lures he amassed in his travels. No country store, mom and pop bait shop, flea market, junk yard, estate sale, hardware store—or fishing buddy's tackle box—was safe from his prying eyes.

Most of the junk Herb bought by the box full was just that: useless trash that had been gathering dust and growing mold in someone's basement for decades. He never could bring himself to throw any of it away, however, because there was always a piece or two he could rob from one lure to bring another back into bass-catching condition. These relics dangled by the hundreds along the walls of his shop and den.

Amazingly, he knew every one by brand name and could pluck it from inventory with a minimum of searching. If I happened to be visiting his place and mentioned a weird plug I'd seen a few days earlier in the cob-webbed corner of some sporting goods store, Herb would immediately walk to the wall, grab a bait off the net where it hung, and say something on the order of, "Did it look like this?"

"Why, I don't remember exactly," I would reply. "I just glanced at it."

That was never good enough for Herb. Invariably, he'd think I had stumbled upon some priceless angling antique from King Tut's tomb and would begin quizzing me like a federal agent.

"What kind of eyes did it have? Were they glass? What color? What condition was the paint? Was the original box still with it? Did the lip have a bend in it or was it straight?"

I had a standard reply for each query—"I don't know"—which would always elicit a stern look and the ten-thousandth lecture about being more observant.

But Herb's true genius with lures had little to do with collecting. He was an originator. I am convinced that, in another time and place, he could have become one of the nation's foremost lure inventors. The guy had a sixth-sense knack about these things.

Some of his creations were novelties. Some were honest-to-gosh lookalikes of baitfish. All of them were capable of catching bass. For instance, Herb showed up one day with a pecan fitted fore and aft with prop blades and treble hooks.

"This oughta drive 'em nuts," he announced. "*Herr-herr-herr-herr!* Nuts, get it?"

It wasn't a world-beater, but by golly, he fished with that pecan long enough to catch a small bass. That was to prove to me, he said, that *any*thing, if properly crafted and presented, would fool a fish.

Apparently this gift runs in all inventive anglers. A little over one year after Herb died, another legendary Tennessee fisherman and tackle innovator, Eddy George, passed away. At the memorial service, fisheries biologist David Bishop told the congregation about once watching Eddy carve a lure from a sweet potato and then use it to catch a bass. What a humbling testimony to the multi-billion-dollar fishing tackle industry!

Another of Herb's more whimsical creations was his "utensil" collection. He took a complete place setting of cheap flatware he had purchased at a flea market and crafted each piece into a functional lure.

"Anybody can fish with a spoon," he told me with a laugh. "I've decided to try the fork and knife, too."

I don't think Herb ever caught a fish on his fork or knife—at least he never did in my presence—but I watched him take a six-pound, eight-ounce farm pond largemouth on his uniquely designed spoon.

Instead of drilling the holes for hook and line on the narrow ends of the spoon, as in conventional models, Herb positioned them across from each other, at the widest part. This created a spoon that wobbled violently as it was being retrieved. Unfortunately, it also wobbled erratically. Herb was never able to fine-tune this invention to run straight on every retrieve, and he eventually quit tinkering with it.

But many of his other homemade lures were classics. With a bit of luck, engineering advice, and marketing moxie, I'll guarantee they could have been mass-produced and still be selling just as briskly, and catching bass just as successfully, as when the prototypes emerged from his shop more than thirty years ago.

One was the "Baby Ray."

It wasn't so much the design of this plug that made it great. It was the color. Herb started with an old Heddon Tadpolly crankbait and set out to reproduce on it the exact replica of a well-chewed crawfish he had extracted from the stomach of a pre-spawn bass. He experimented with a variety of colors before settling on a pattern that can only be described as an "off-white, yellow-brown" background streaked with beige and tan.

Yes, it was just as ugly as it sounds. Instead of "Baby Ray," I called this grotesque beast the "Baby Puke." But in Herb's deft hands, it was an absolute killer on early-season bass in shallow, muddy water. Of course, he only carried one in his tackle box at any given time.

Another Hubbard original was the "Slugger."

Herb came home from the flea market one day with a couple-dozen miniature baseball bats, each approximately four inches in length. To the best of my memory, he said they were part of a toy collection. Herb was not interested in toys. In his eyes, these light-colored pieces of wood looked like a minnow.

He attached props and treble hooks to either end of one bat, painted eyes and gills, and premiered it during one of our float trips down Little River, a rushing stream that surges out of the Great Smoky Mountains National Park.

I'll never forget the moment he produced it from his tackle box. It was shortly after we had beached the john-boat and eaten lunch. We shoved back into the current and he announced, "It's obvious that if we catch any decent fish today, I'm gonna have to do it. I happen to know what these bass need."

"Such as?"

"They need to be beaten in the head with this bat," Herb answered, proudly attaching his Slugger and holding it aloft for my inspection. "Whadaya think?"

"I think you're nuts. What fish in its right mind is gonna mess with a ball bat?"

"Oh, ye of little faith," he replied. "Sometimes, I don't know why I even fool around with tryin' to teach you anything."

It took Herb roughly ten casts with his baseball bat to detonate four pounds of Little River dynamite: a classic, cold-water, swift-current, leap-for-the-sky smallmouth bass. It was the first of six smallmouths he extracted during the remainder of the float.

Do you even have to ask if there was more than one Slugger in his tackle box?

"I'd let you use this one," he said, doe-eyed and dripping with bogus sympathy, "but like you said, no fish in its right mind would mess with a ball bat. *Herr-herr-herr-herr!*"

Herbie's mind was so fertile, it could crank out on-the-spot "originals" having nothing to do with hooks and spinner blades. Once when he, his son Major, and I were on Fort Loudoun and the shear pin on our outboard motor broke, he used the straight piece off of a belt buckle for repairs.

Not surprisingly, it was my belt buckle.

"Your britches are tight enough as it is," Herb observed. "They don't need a belt to stay up. How did you gain that much weight in the last few months?"

Before I could think of a protest to set the record straight, he quipped, "I know you haven't been catchin' enough fish to stink up a skillet! *Herr-herr-herr-herr!*"

Another time, he showed up at my office with a new, double-bladed buzzbait.

"Come see how this thing works," he said, excitedly. "It makes more noise than a six-horse Johnson."

"Where are we gonna try out a fishin' lure in downtown Knoxville?" I asked.

Herb glanced out the window. "C'mon," he said, grabbing me by the sleeve.

We walked down the hill to a gravel parking lot. It had rained the day before. There was a huge mud puddle, about the size and shape of a kiddie swimming pool, in the center of it. He stood there, among the starlings and the strangers, casting that buzzbait and marveling at its sound, just as if we were in the wilds of Canada. The only thing that surprised me was the fact that a three-pound largemouth didn't materialize from the depths of that puddle and engulf his new creation.

Herb's inventive streak even transcended the world of fishing. One time—don't read this if you have a fear of dentistry—he repaired his broken dentures with Super Glue, but not before filing down the sharp edge of one of his real teeth—*arrrgh!*—with an emery board. The term "don't try this at home" was not in his vocabulary.

Yet of all of Herbie's originals, none could hold a flame to his "skin plugs." Once again, he was ahead of the curve.

In the late 1970s, Madison Avenue discovered the word "natural." It became an integral part of the commercial message for every conceivable product. The most nutritious breakfast cereals were "natural." The finest facial soaps were "natural." The most fashionable shirts and dresses were "natural." I didn't make a survey at the time, but it wouldn't have surprised me if, through some sort of commercialized alchemy, artificial sweeteners and synthetic fibers were hawked as "natural," too.

It wasn't long before the nation's fishing tackle manufacturers embraced this trend. Virtually every national company introduced a line of "naturalized" lures painted to imitate nature, right down to a premium scale finish.

It was old hat to Herb. Half a decade earlier, he had come to the conclusion that Mother Nature's color patterns needed no improving, nor could they be accurately duplicated with paints and flashy foil finishes.

Why try to duplicate the look of a forage fish on an artificial lure? Why not transfer the actual baitfish itself? If countless thousands of generations of bass had been gorging on prey species for lo these many years, why not shoplift a link from the food chain?

That's when Herb started making plugs out of skins and scales. I am not making this up. I have sat in the boat and watched him do it. Several times.

He would begin with a plain, flat-sided wooden plug he had whittled himself out of poplar. Then he would catch a small baitfish, usually a bream or pint-sized bass. He would kill the victim with a blow to the head and, in about the time it takes to type this sentence, remove its skin with the tip of a razor-sharp fillet knife.

A couple of minutes in the sunshine would dry the skin just right—not too pliable, not too stiff. With scissors, he would cut a swatch of skin to fit either side of the plug. Then he would fix it in place with fast-drying glue and start casting.

I wish I had timed the entire process but never did. Thinking back, it seems hardly half an hour elapsed until the retrofitted minnow was back in the water, luring bass to their demise.

The skin plugs Herb crafted on-site were pretty crude, particularly the early ones. As he got better and faster at the process, they began to look more lifelike. He started taking fish skins back to his shop and, after gluing them in place, coating the entire plug with a thick, clear lacquer finish.

These were such works of art, I was never comfortable fishing with them. Chipping a four-dollar store-bought plug against the rocks is forgivable. Banging one of Herb's heirlooms would be a sin beyond redemption.

In my office at home, I still have two of these jewels. Neither has ever been near the water. They never will, either, even though they would surely prove deadly.

Best I can tell, one of them is a combo: part shad, part bluegill. The body reflects the silvery side of a shad, but there's also a gill flap from a male bream.

The other is solid gold, figuratively speaking. The skin came from a large ornamental goldfish that formerly swam in my wife's garden pond. No, Herb didn't sneak in one night and make off with it. But every time he visited our home, he inspected the pond and took inventory.

"If that goldfish ever dies, I want it," he said.

Finally, old age took its toll—Herb swore to his dying day he did nothing to hasten the deed—and I froze the dear departed until I could deliver its carcass to the undertaker. A couple of weeks later, Herb showed up at the front door with the reincarnation.

"I didn't put hooks on it for a reason," he said. "It wouldn't be fittin' and proper to go around catchin' bass on poor ol' Mary Ann's dead goldfish. So I made a 'mee-morial' for her to remember it by."

Herb's gold-colored, hookless "mee-morial" still rests on the bookshelf. I've never been so much as tempted to fish with it.

But I'll bet somewhere, deep in the bowels of one of his storage closets, there's a duplicate gold plug, complete with hooks, that has seen plenty of duty. Maybe a couple of them, for that matter.

You see, the dead goldfish I solemnly delivered to Herb's house was a good seven inches in length. That's a total of fourteen inches of potential skin for making lures, less perhaps an inch or two of waste around the head, eyes, and dorsal fin. Yet only about four inches went into the making of the "mee-morial."

And like I said before, Herb never threw *any*thing away.

"Murph," a.k.a. "Queen Mary"

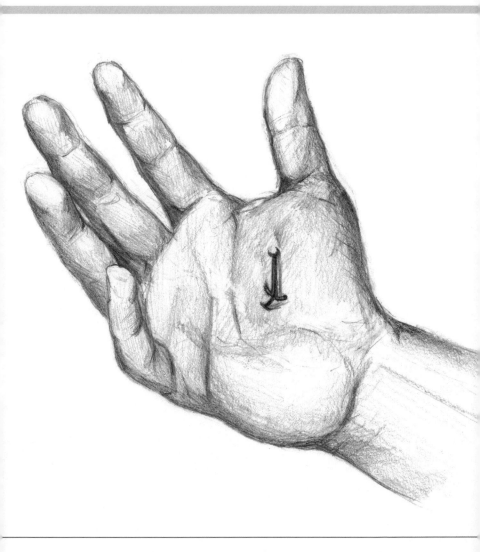

T he January 2000 issue of *Bassmaster* magazine contains the results of a field test on the Stratos 19 SS Extreme bass boat. The report appears on page 46. I am looking at a copy of it as I type these words. Fishing fanatic Don Cruze, one of my former colleagues at the *News-Sentinel*, clipped the story out of the magazine and mailed it to me.

"You need a boat like this," Don wrote in an accompanying letter. "I'm sure it would make you a much better bass fisherman."

I read the article and immediately reached the conclusion that Don was dead wrong. The only thing a boat like this would make me is broke.

The test boat came equipped with a two-hundred-horsepower outboard engine. It generated a top speed of 72.6 miles per hour. The list price, as rigged, was $38,814, including options like a heated driver's seat and a liquid crystal tachometer-thermometer on the dashboard.

I am certain the Stratos 19 SS Extreme is a marvelous piece of nautical workmanship. Judging from the color photograph, it could win all manner of beauty contests. Bully for the people who buy one.

I don't need that much boat, however. Never have, never will.

When Herbie and I entered into co-ownership of a float-fishing boat in 1980, we settled for something substantially less. We chipped in sixty-five dollars each for a second-hand, much-dented, eleven-foot, eight-inch aluminum johnboat. It became the symbol of our dichotomous relationship.

Herb named it "Queen Mary" in honor of his wife, Mary Jo, and my wife, Mary Ann. I named it "Murph" after Murphy's Law.

He always fished, and drank iced tea, from the back. I stayed up front—on a seat marked "Pole Position," naturally—with my barley juice. There was an imaginary line drawn across the boat, exactly halfway between bow and stern. Anytime I had business in the rear of the vessel, I was under orders to come alone. The barley juice was not welcome.

With the addition of Murph to our flotilla, no body of water was safe. If the attack orders called for an assault on the main channel of a major reservoir, we had either of our big boats. If the battle plan switched to farm ponds and shallow bayous, there was the Hub Tub. And for float trips down Little River, the Holston River, the French Broad River, and other flowing fisheries, Murph was summoned. In all the navies of the world, no two admirals have ever been blessed with such vast resources at their disposal.

Murph was narrow enough to slide easily into the bed of my pickup truck or the back of Herb's van, yet wide enough for comfort and safety on the river. It was light enough for easy portaging, yet heavy enough to bounce harmlessly, albeit somewhat more dented than before, off of mid-stream rocks.

No boat under twelve feet in length could ever be deemed spacious, but Herb and I never felt crowded going downriver in Murph, despite the inclusion of rods and reels, cameras, tackle boxes, a battery for the electric motor, and coolers. After working for so long from the confines of the eight-foot Hub Tub, expanding our horizons to eleven feet, eight inches was like moving from a cramped studio apartment into a three-bedroom rancher.

Before Herb's failing health began robbing him of endurance, floating for largemouth, smallmouth, and spotted bass in Little River became our passion.

Float trips require a flexible schedule, plus two vehicles—one for "put-in," the other for "take-out." The complicated drill of dropping off the take-out car downstream, ferrying everything upstream to the put-in site, making sure lunch, gear, sunscreen—and, most important, keys to the take-out car—are not locked securely inside the put-in car (trust me; it's happened), and then reversing the process ten hours later, makes heart surgery appear spontaneous.

These are dawn-to-dark affairs. Once you shove off Point A, headed for Point Z, it's downhill all the way. There's no going back upstream.

Except under extenuating circumstances. . . .

"You're the only person I ever saw who could hook himself tryin' to unhook a fish," Herb was saying. "How in the world did you manage to do that?"

I wish I knew. Perhaps it was simply time for my number to come up. After a lifetime of fishing, I had been nicked, pricked, zapped, and dinged by hooks on a number of occasions. But the pickle I was in right now was more serious. I had managed to drive one section of a large treble hook not only into the palm of my right hand, but then up and under the "drumstick" of my thumb. We're talking buried to the hilt.

Herb and I were fishing that day with Lowell Branham, a longtime friend, *News-Sentinel* copyeditor, and outdoor columnist for the Scripps Howard News Service. Mercifully, the three of us were not crammed into Murph. Lowell had joined the armada in his canoe. We had launched both boats at Davis Ford, near the Walland community, and were going to float several miles to the old Cave Roller Mill at Wildwood.

About half a mile downstream from the launch site, we beached our boats at the head of a long run. That way, we could wade the shoal and slowly work every bit of the pool beneath it.

On my third or fourth cast, a ten-inch spotted bass inhaled my crankbait. It was a true piscatorial pipsqueak.

But never let it be said that a pipsqueak can't whip a full-grown man.

In truth, carelessness caused the whole thing. I was too busy worrying about the next cast and wasn't paying close enough attention to the immediate chore. I swung the fish free of the current and reached out to catch it like a baseball.

When the bass hit against my outstretched hand, it shook violently. It just so happened that one exposed point of the rear treble hook hit my palm at the same instant. Before I could yell "rock-elephant!" (or words to that effect) at the top of my lungs, the deed was done.

Instinctively, I tried to drop the fish. That was a big mistake.

The bass, now dangling from my hand, began to beat wildly, driving the barb further in with every thrash of its tail. It was a scene which, if captured on tape, surely would have won first prize on one of those "funny home video" television programs.

Most assuredly, there was nothing funny about it at the time. I threw down my rod and began grabbing at the bass with my free hand. I finally nabbed it. But the rock-elephant quotient, now at an all-time high, was about to go off the charts.

Lowell and Herb waded over to assist. It was obvious surgery would be required—with pliers on the lure and with a knife on my hand. Both men reached into their respective shirt pockets for their "cheaters" (reading glasses). Simultaneously, they discovered both pairs were safely locked in the put-in vehicle.

"Aw, I don't need those glasses anyhow," said Herb. "Gimme your hand."

Securing the still-squirming bass in place with my left hand, I pushed my right hand in Herb's direction. He squinted and pulled his head back, the way folks do when they can't read the fine print of a want-ad. Then he spoke the words no patient ever wants to hear from a health-care provider.

"Wait just a second and let me get ever'thing in focus."

Herb produced a pair of pliers. He turned them flat and positioned the side-cutters on top of the O-ring attached to the eye of the hook. Lowell, using his fingertips, held the plug in place.

Herb bore down forcefully on the handles, and I rock-elephanted the wax out of his closest ear.

"When's the last time you sharpened that thing?" I hollered.

Herb ignored my squalling.

"Hold still, you big baby! How d'you expect me to operate when you're squirmin' like a worm on a hot rock?"

"Baby's butt! You're gougin' hunks of meat outta me!"

It took four attempts, but, amazingly, the blind mice finally managed to sever the connection between bass and man. Back into the river went the fish. Now, there was this small matter of the hook still remaining in my hand.

This was not your garden-variety impaling. The metal was buried beyond the bend. Only a portion of the shank was exposed. The three of us agreed it was a job for professionals. I cast the most vocal vote of the electoral process.

"Ray, you stay here," said Lowell. "Me'n Sam'll take my canoe and paddle back upstream to his truck. Then we can run over to Blount Memorial Hospital's emergency room. With any luck, we should be back in a couple of hours."

Indeed, we were back inside two hours. But not before enduring another run of bad luck.

That's because Lowell, sadist that he is, paddles and fishes from a keel-less canoe. Safely maneuvering one of these nautical banana peels is difficult enough in calm water. Paddling against the current, with a bum hand, only added to the pleasure.

As expected, we deep-sixed—fifty yards from the truck, of course—in the deepest hole of the entire run.

Spitting water, Lowell and I dog-paddled, swam, and waded the flooded canoe into the shallows, emptied it, and buried it in streamside brush. We climbed into my truck and drove to the hospital. Lowell stayed in the parking lot while I sloshed—mud literally was squirting out of my shoes with each step—into the emergency room.

The nurse looked up in horror, not so much at my plight but at the condition of her floor.

I held my hand aloft and said, "I got a problem."

Naturally, there wasn't the first sign of an insurance card in my waterlogged billfold. It was at home, safely tucked in a drawer so it wouldn't get lost. As my hand was being prepped for the plucking, the nurse was kind enough to call my office for the necessary identification numbers.

The attending physician gave me good news: "You were wise not to attempt getting this thing out by yourself. There's a tendon wrapped around the barb. If you had tried pulling it out, we could have heard you from all the way over at the river."

And then the bad: "I'm not gonna lie to you. This is gonna hurt like hell for a couple of seconds."

With that, he produced a syringe the size of a putty gun with a screwdriver sticking out one end of it. At least it looked that big to me.

He soused down.

Doc, bless his rock-elephant heart, certainly was telling the truth. For the first couple of seconds, it felt like every hornet in Blount County was stinging my thumb. Then, just as quickly, my entire hand went numb. I believe he could have chain-sawed me at the wrist, and I wouldn't have felt a thing.

"It's probably useless for me to tell you not to go back fishing today, isn't it?" he asked after the ordeal was over.

"It sure is," I replied. "All my gear's on the river, along with another buddy. I've got to float on down to the other truck, anyway, so I might as well fish."

"Well, at least try to keep it dry."

"I'll do my best."

We both knew that was a lie. Float-fishing involves more than a bit of moisture. Between splashy trips through whitewater and portages around the more serious shoals, nothing escapes an occasional dousing.

Doc said he understood. He fished the river, too. But he had to tell me to keep the wound dry, anyway. Must be something about it in the Hippocratic Oath.

Miraculously, Lowell and I managed to get back downstream to Herb without incident. We considered it a good omen for the remainder of the journey. This was an incorrect assumption.

"What'd you all do—go to Knoxville for lunch?" Herb hollered as we paddled into view. "I'll swan, it's awful hard to go fishin' with a bunch of fellers who lollygag around town all mornin'."

"Speakin' of lunch, I think I'll have somethin' to eat," I said. "Where's that sausage biscuit left over from breakfast?"

An angelic countenance swept over Herbie's face.

"Oohh, did yeeew want that?" he said. "I figured you were hurtin' too much to eat. Boy, it sure was good. I wish you could have tasted it. *Herr-herr-herr-herr!*"

Without thinking, I reached into the river and splashed him with a handful of water.

Oops, wrong hand.

Oh, well. Got that out of the way. For the rest of the trip, I wouldn't have to worry about keeping the dressing dry.

We shoved off and resumed our mission. Despite the interruption, this turned out to be a banner day as far as the bass-catching was concerned. In hardly over a mile, we boated eight beautiful smallmouths, the largest a little under three pounds. Most were caught on top-water lures. Until you have gone surface-to-surface with a square-shouldered river smallmouth bass, its muscles honed by a lifetime in the current, you've never savored one of fresh-water fishing's most exciting moments.

Casting to them, however, was another matter.

Even bandaged, my hand was serviceable. All the fingers worked on command. It's just that the muscles stayed numb for the first couple of hours, making the sensation of "thumbing" the line on my casting reel a surreal experience. I managed to send more than one lure into the trees.

Herbie was completely sympathetic to my plight: "You fishin' for smallmouths or squirrels?"

Like our other boats, Murph did not have a livewell—unless you take into account the inch or two of water that always welled up in the bottom during river missions. Thus, we deployed a rope stringer for fish-holding purposes. One crucial duty on float trips is to pull the stringer in every time the boat bumps and grinds over rocks. Early on, Herb decided this was a job for the man in Pole Position.

At the first shoal below the Highway 411 bridge, the man in Pole Position was far more concerned about fishing than stringer procurement. Just as Murph cleared the whitewater and the man in Pole Position cocked his arm to cast, there was a violent jerk. Murph swung sharply in the current, stopped momentarily, then suddenly began drifting again.

The man in Pole Position and the man in the back seat both looked overboard in time to see the remnants of a rope stringer drift into the depths of the clear water and a covey of freshly liberated smallmouth bass explode like quail. That's about as close to "rock-elephant" as Herb ever came.

"And to think I was gonna fry up a big mess of those fish for you and Lowell and me when we got home," he sighed. "Had it all planned out."

"When did you ever fix fish right after a trip?" I shot back. "You always claimed you were too tired. If I'm not mistaken, you were even too tired to clean them or put away the gear."

"Oh, this time I wasn't gonna be tired. I got in a *gooood* rest this mornin' while you and Lowell were goofin' off in town."

Politicians and comedians alike should have tutored under Herb. He was never at a loss for a comeback.

We practiced serious catch-and-release the remainder of the trip. Which is just as well. On river trips, fish are a bonus. Even though you are usually sun-baked and dog-tired by the time you reach the take-out vehicle, these are the ultimate in relaxing outdoor experiences. No studio recording could ever completely duplicate the background symphony of songbirds and babbling waters. Anybody who comes off a river float with high blood pressure just ain't human.

Except maybe this time.

The pool above the Wildwood Bridge is loaded with redeyes, a scrappy, dark-skinned panfish a bit larger than a bluegill. Herb was a master at catching them on tiny crankbaits. Because of their light weight, however, these mini-lures required more arm-sweep than wrist-flip when presented on conventional casting equipment. The demand for wide berth was accentuated by the necessity of laying the lure beneath the limbs of overhanging trees.

It just so happened, right before one of these casts, that Murph had spun around in the current. Or maybe Herb had pivoted us with Mrs. Reed when I wasn't looking. In any event, the man in Pole Position was still sitting on his seat, with his barley juice. But the man in the back seat, with his iced tea, was suddenly up front. Thus, when the iced tea man leaned way over to cast under the trees, the arc of his rod came sweeping across the opposite end of the boat. Like a guillotine.

And it just so happened that the man in Pole Position was deep in concentration, slowly working a plastic crawfish along the bottom of the river. Thus, he was holding his rod aloft, in a vertical position. Like an antenna.

Ker-sworp! Snap! Fuzzzz!

The top six inches of my rod were cleanly lopped off. Herb couldn't have done a better job if he'd used hedge clippers. But the sudden impact—anybody who's ever used a casting outfit knows what I'm talking about—had backlashed his reel worse than a rat's nest.

"Now just look what you've done to my reel!" he exclaimed.

"Your reel's ass!" I shouted back. "Look what *you've* done to my *rod!*"

Herbie didn't miss a beat: "If you wouldn't use such cheap equipment, it'dud hold up to a little tap. Don't those rod makers send you better stuff than that? Why, you oughta be ashamed of yourself!"

Herb surveyed the damages and made immediate contingency plans. Since his reel was hopelessly out of commission and my rod was busted, it only made sense to put my reel on his rod for the rest of the trip. And since we were so close to the truck, there would only be time for one man to fish effectively.

He gave me Mrs. Reed as he reached for my reel.

"Make yourself useful," he said. "See if you can keep us straight in the current. *Herr-herr-herr-herr!*"

Years after the fact, the man in Pole Position, now sitting in his office and typing these recollections, just now did the same thing he did that afternoon, right along with Herb, when the day of disasters was about to come to a close.

He busted out laughing. If only to keep from crying.

I've still got Murph. It rests, upside down, in a place of honor beneath my front porch. When we built this house, I had a concrete pad poured under the porch, just so Murph could stay high and dry off the ground. Occasionally, I drag the old war horse out and use it when I'm duck hunting or fishing on farm ponds. But it just isn't the same.

No, our $130 river boat wasn't fancy. Unlike the Stratos 19 SS Extreme, the only time Murph had heated seats was when Herb and I were fishing in direct sunshine. Unlike the Stratos 19 SS Extreme, the only way Murph would have achieved a top speed of 72.6 mph was if (a) it had been lashed to the bed of my pickup truck and (b) I had driven off a cliff.

But as for the Stratos 19 SS Extreme's price tag of $38,814?

That wouldn't touch the good times we shared in Murph. Wouldn't even come close.

Barney and
His Bullet

It didn't take too many ventures with Herb to realize just what a master of efficiency he was.

Despite the fact he owned enough lures to stock a dozen sporting goods stores, he rarely went to the lake with more than eight or ten plugs at a time, usually carefully laid out in one of those old cigar box tackle carriers or the gold patent-leather purse. The man knew what he was fishing for, he knew what types of conditions he would encounter, and he knew what baits would be needed to accomplish the task. So why burden the process with non-essentials?

One Thanksgiving Day, I discovered this principle applied to hunting as well as fishing.

We were headed, three pickup truckloads of us, to Amos Stafford's farm in neighboring Loudon County for a morning with rabbits. "We" covered a multitude of hunters, both two- and four-legged. There was Herb and his two sons, Major and David, plus one of Major's friends, Charles Purkey, as well as a trio of Major's eager beagles, Snoopy, Susie, and Dixie.

The first order of business was to stop at a grocery store long enough for Herb to purchase a "host gift." At least that's what it is called in etiquette books. It's usually a bouquet of fresh flowers, a loaf of exotic bread from some gourmet bakery, a bottle of wine—anything to serve as a token of esteem and goodwill.

There is no need for good ol' boys to consult etiquette books.

"Flares" are what grow in women's gardens. "Lightbread" tastes decidedly better with baloney than that rockhard stuff made with sun-dried tomatoes. And anybody who'd dare suggest wine, particularly in the company of a straight-laced country preacher, is off his rocker.

Herb came out of the store with a three-pound can of JFG coffee which, I would be willing to bet a week's wages, would be more appreciated by any host, any time, any place, than all the posies, pastries, and powerful potions ever sold in fancy shoppes.

When we reached the Stafford place, Amos was on the front porch, lacing up his boots. He thanked Herb for the coffee and insisted we all have a cup. By the time we finished, the hounds were verily singing in their box in the back of Major's truck. They knew this was opening day of the rabbit season in the South and, by golly, it was time for action!

Hunting dogs can sense these things. I am certain of it. I've seen it too many times. I have been around hunting dogs all my life and have raised an assortment of them myself, from hounds to retrievers to setters and pointers. Any dog that doesn't start jumping and yapping at the sight of shotguns and hunting duds—grizzled-faced veterans are permitted to merely wag their tails and smile—is not worth a fifty-pound sack of Purina. As the guns were being uncased, Snoopy, Susie, and Dixie started hitting the high notes.

Major opened the door of the box and was nearly trampled.

We fanned out into the first field. As we did, I glanced at the assembled artillery in amazement. There was enough firepower in this crowd to wage minor warfare. Charles, Amos, David, and Major were feeding high-powered sixes into twelve-gauge automatics. I was doing the same with a sixteen-gauge pump.

Herb, on the other hand, had opted for substantially less armament. He was carrying a .22 caliber rifle. Unloaded.

"How do you expect to hit a rabbit that's burnin' up the ground with that pea shooter?" I teased.

"Oh, I'm just along to hear the dogs work," he answered. "You boys can do all the shootin'. I can't afford all that powder and shot. Besides, those big ol' shotguns are too heavy for a poor ol' man like me to tote all day. Me'n this little rifle will get along just fine."

One of the beauties of hunting in the company of the landowner is that you don't cover a lot of unnecessary territory. Amos was intimately familiar with the 150 acres of his farm, and he knew the direct route to the most productive rabbit cover.

Since he raised beef cattle, part of his acreage was committed to fescue—not the ideal habitat for bunnies, birds, and other small game, but Amos had let plenty of edges, corners, gullies, and swales grow up for the critters. We crossed the hill and immediately found ourselves in a wildlife haven, choked with briars, honeysuckle, and sage grass.

Amos kicked a brush pile and said, "There's always a few rabbits aroun—*there he goes!*"

The cottontail bounded through the brambles like a racehorse. It was gone before anyone could so much as push the safety, let alone shoulder a gun, aim, track, and fire.

"Yo-yo, dogs!" Major hollered. "C'mere, you all! Yo-yo-yoooo!"

The beagles needed no further instructions. Three noses hit the ground, and the music began.

Hunting and fishing lend themselves to a variety of pleasing sounds, and there's no better stage for the performance. A hearing-impaired person can enjoy these sports—indeed, many do—but I cannot imagine the handicap they face. Without the honking of geese as wind rattles the dried cornstalks in January, or the thunderous gobbling of a wild turkey on an April morning, or the slurping "kiss" as a fish feeds on the surface, the experience would surely be halved.

But I'm not so certain that the yodeling of beagles can't top them all.

This is not necessarily an excited bark, like coon hounds at the tree. It's more of a melodious cry that varies in pitch depending on the degree of intensity. Cold-trailing beagles occasionally emit a frenzied wheeze, if for no other reason than to convince themselves to stick with the game plan.

Beagles that have temporarily lost the trail whine and bark impulsively, their whipping tails all but a blur as they try to unravel the scent. But a pack of rabbit dogs bawling all-out on hot spoor is a concert.

This would have been show enough if the beagles had concentrated on the rabbit Amos jumped. But before they could course the first bunny more than a few hundred yards, the dogs kicked up one of their own. Two rabbits for three beagles is too much of a good thing. Instead of ballooning the dividends, it sends the stock crashing.

Snoopy couldn't convince Susie and Dixie to stay focused on the primary mission. Or maybe it was the other way around. Whatever the circumstances, the rabbits tied the trio in knots and scampered to the safety of their burrows.

"Don't worry," said Amos. "There's plenty more where those came from."

You better believe it. Before the morning was over and we parted company to join our own families for Thanksgiving Day feasts, nine cottontails were in the bag. Six of them came courtesy of the beagles' ability to trail their scampering quarry across the autumn countryside and bring it back toward the shotgun army. The three others came courtesy of Herb's .22

He shot them, sitting—*spat!*—with a single round each, right behind the ear. He never even bothered to feed a cartridge into the chamber until he had spotted his target.

I've seen other old-timers hunt rabbits this way. There is nothing whatsoever unsporting about the method. As far I'm concerned, this is the ultimate in careful, leaf-by-twig-by-stem-by-branch inspection and then stalking. It is the epitome in predator-versus-prey. Hubert Faulkner, the man who introduced me to rabbits and beagles when I was ten years old, was a virtuoso in this regard.

"Just look for their eye in the brush, Sammy," Hubert had instructed in those bygone years. "It's a shiny brown dot. You ain't ever gonna see the rabbit hisself. That's why he's got that fur coat. It blends in with ever'thin' around him. But he'll always have his eye open, a'watchin' you. That brown dot'll give him away ever' time."

That's precisely what Herb was doing.

I watched him comb one section of the field, away from the main hunting party. He hardly moved at all. Just a step every eight or ten seconds as his eyes scoured the cover. It was more like watching a red-tailed hawk, sitting atop a dead snag, inspecting the ground below.

Only when he spied a rabbit would Herb reach into his shirt pocket, fish out a .22 short, load the gun, and—*spat!*—produce one more for the table. No muss. No fuss.

We gathered back at the house for a goodbye cup of host coffee, divided up the plunder, and headed for home.

"I really don't want but one rabbit to eat," Herb told me as I pulled out of Amos's driveway. "What have we got back there—three or four?"

"Yeah, somethin' like that," I said.

"Well, you take all but one."

I knew what was coming next.

"And since you're gonna be doin' a bunch of skinnin' anyhow, why don't you just skin that one for me before you go home?"

I wouldn't have had it any other way. Nor did I need further instructions, even though I felt certain they were about to be issued.

"Make sure the one you skin for me is one of those I shot in the head," Herb said. "I don't want to spend half my meal spittin' out shotgun pellets. Now, do you reckon you could take these curves a little bit slower? All that huntin' has wore me out. I need to get a little rest before dinner."

With that, Herb's head went back on the seat. His lights went out. He began snoring—first heavy breathing, then wheezing, then full-blown bellowing.

It sounded like Snoopy, Susie, and Dixie as they deciphered a cold trail and kicked the race into high gear.

Little White Lies

"T he way you tell lies to me," Herb preached, "is an absolute scandal!"

He and I were standing outside an old country store, just off Highway 411 near the Knox-Blount county line, when he delivered that sermon. I thought about it the other day when I drove by the building, which is long-since closed down and boarded up. Even now, more than twenty years after the fact, I could still hear his words hanging in the cool of that late-August evening.

"I'll swan, when you die and go to heaven—assumin' you make the grade—they ain't gonna have just a page of your sins to discuss. No, buddy! You've racked up so many lies, partic'arly to me, it's gonna take a window shade to hold 'em all. You'll start to plead your case, and ol' Saint Peter'll just yank down that window shade, and I reckon they'll vote you out, right there on the spot."

"You'll come argue on my behalf, won't you?" I asked.

"Highly unlikely," he replied. "That might not be the best place to get caught associatin' with the likes of you. Wouldn't be good for my image."

"Wha'daya mean your image? You're the one who's taught me how to tell lies. Fishin' lies, anyhow."

"I have *not* taught you how to tell lies," Herb countered. "I have taught you—or tried to teach you—to not go around blabberin' ever'thing you know. There's a distinct difference."

"What kind of difference?"

"Lies—the sort of stuff you tell—are things that aren't true. Not blabberin' ever'thing you know is tellin' the truth, but just not too much of it at any given time.

"It's just like the other day, when we were pullin' the boat out of the water at Unitia. We had us a good string of bass— no thanks to you, of course—and they were in the bottom of the boat, down there where nobody could see 'em. There's no sense advertisin' your fish. If you do, ever'body for half a mile around will want to know where you caught 'em and what you caught 'em on. Then the next time you go, the place will be so crowded you'll have to quit fishin' for a month. But what did you do? You had to start blabberin'."

"I was just being polite," I said. "All I did was answer a question for those other fellows. They asked if we had done any good, and I said, 'Yeah, we didn't do too bad at all.' That was simply tellin' the truth."

"No, it wasn't tellin' the truth," said Herb. "First off, it was an outright lie because you hadn't done that much good. If I remember correctly, I'm the one who caught most of those bass. But then you made things worse by showin' the fish off."

"That's where *you're* tellin' a lie," I spoke. "I didn't get a chance to show 'em the fish 'cause you jumped in the truck and said to quit talkin' and for us to get goin'. I bet those fellows thought we were awfully rude."

"You'da thought they were awfully rude if you went back there in a few days and they were fishin' your spot, wouldn't you?"

"Herbie, Fort Loudoun Lake covers thousands of acres! How is somebody else gonna figure out where we were fishin'?"

"I don't know how they'd do it," he grumbled. "They just would. And it sure makes it easier for them to do it when there's some blabbermouth givin' out directions. Why, I'm amazed you didn't want to stop by a radio station on the way home and broadcast it to the entire city!"

"So what you're sayin' is that honesty ain't always the best policy?"

"No, that's *not* what I'm sayin'. Honesty is always the best policy, just as long as you don't dish out too much of it at once. Anytime you've got some fish, don't be showin' them off—at least not if you're near the water. It's OK once you get home. But for Pete's sake, anytime you're at the lake or river, stay quiet. Just drop the fish in the grass, if you have to. The least you can do to keep from attractin' a crowd, the better."

I was toying with Herb, as usual, baiting him further into an argument. It was a game we loved to play. But it was a game I knew I'd never win. Herb always had the last word. This time. Every time.

Actually, I did feed him unvarnished lies on a regular basis. I didn't do it out of meanness. Not at all. I did it because he was so gullible. The more outrageous a lie I could dream up, the deeper Herb would swallow it. For instance:

I telephoned his house one afternoon, three days after he and I fished the Stock Creek embayment of Fort Loudoun in the Hub Tub. It just so happened that I was attempting to diet at that time—a rare event, invariably ending in failure and two new pounds. Herb had showed up with half of a chocolate cake that Mary Jo had baked. He couldn't believe that I wouldn't take so much as one delectable bite out of it.

An official-looking car pulled into the ramp area while we argued about the cake and launched the Hub Tub. It probably was something as tame as a county detective stopping to radio back into headquarters for directions to a crime scene. But no opportunity for panic was lost on Herb. He was convinced somebody was spying on us.

Thus, I could almost hear him gulp that afternoon when, after he answered the phone, I said in my best disguised voice, "Mister Hubbard, this is Commander Jones from the United States Coast Guard."

"Yes, sir," he replied, "what can I do for you?"

"You own a small boat, don't you?"

"Well, it's kindly small."

"And the numbers on it are 'TN 8259 AA?'"

"That might be them. I don't remember."

"Mister Hubbard, did you have that boat in the Stock Creek area this past Tuesday?"

"Well, I really can't remember."

"Mister Hubbard, one of our agents was driving by and saw this boat being occupied in an unsafe manner."

I could tell Herb was getting steamed. But he was still intimidated.

"What are you talkin' about?" he said. "What kind of unsafe manner?"

"The boat appeared to our agent to be severely over-loaded."

"What?"

"Yes, sir. The man on one side of the boat, which our agent believes is you, was very heavy. The man on the other side of the boat was slim and trim and. . . ."

Then I erred big time. I snickered.

"Who is this?" he thundered.

"It's Commander Smith from the Coast Guard," I answered, fighting to regain my composure.

"I thought you said your name was Jones."

That's when I lost it. I started hoo-haaing and hee-hee-ing. Herb was not amused at first. But belatedly, he started laughing, too. Then he zinged me.

"You'd think a man who lies as much as you do could keep his stories straight!"

Another time, Herb decided to use me as his expert wit-ness. The engine in his van had developed an infrequent miss, causing it to lurch unexpectedly on the highway. As is usually the case in such matters, it purred like a kitten when he drove it to the dealership for repairs. Then when he'd get it back out on the road, it would start jumping like a frog.

"They think I'm stupid down at that shop," he announced one day. "This crazy car won't ever act up when they're lookin' at it. It just waits till I'm alone, then it starts jumpin'."

As if commanded by Detroit, the van suddenly jerked.

"There it is, right then!" he exclaimed. "You felt that, didn't you?"

"Yeah."

"Good. We're goin' over to that dealership right now."

With that, he made a U-turn and sped back down Alcoa Highway. En route to the mechanic, the vehicle must have jumped seven or eight times.

As soon as we pulled into the service bay, Herb grabbed me by the arm and virtually dragged me to the counter.

"You don't have t'take my word for it any more," he said to the technician. "This feller's been ridin' with me, and he can tell you how that van has been jerkin'. Go ahead, Sam."

I screwed up the most quizzical face I could muster and shrugged my shoulders at the technician.

"I don't know what this guy's talkin' about," I told him. "I was just standin' out there in the parking lot, and he came up and said he'd give me five dollars to walk in here and say somethin' bad about his car."

Then I turned to Herb and stuck out my hand.

"Where's my five bucks, buddy?"

If looks could kill, I should have dropped dead right there on the spot. I let him fuss and sputter for a good thirty seconds before I started laughing and 'fessed up to the technician.

And then there was the time at Christmas when I gave him a bottle of liquor for a present. Or so he thought.

Herb loved molasses. He always kept a jar on the kitchen table at home. Knowing his fondness for this sugary goo, I bought several pint jars at the Museum of Appalachia's gift shop. Then I stacked the jars inside of a tall whiskey box and wrapped it in holiday paper.

We met for lunch at the Cherry Pit a few days before Christmas. By that time we were regulars at the restaurant. All the wait staff knew us by name. I gathered everyone around our table after the meal and asked them to witness the grand unveiling.

When Herb tore into the wrapping paper and the words "Old Grand-Dad" came into view, he audibly gasped. I thought the poor man was going to pass out. He tried to shove everything under the table, amid shouts from the assembled masses to, "Crack it open and let's all have a drink!" It wasn't until I rescued the box and tore open the lid to reveal the molasses inside that he agreed to put his gift on public display.

The box remained out of sight, however. And the thirsty wait staff dispersed, quite disappointed.

The country store episode—the one for which I received my sermon about lying—involved a different type of truth-tampering. It occurred late one afternoon after Herb, Dana Keeble, and I had fished a series of Blount County farm ponds.

These little lakes can produce extraordinary results. In fact, this was the day I described in an earlier chapter, the day when Herb caught a six-pound, eight-ounce largemouth on his sideways-running spoon. We visited eight or nine ponds that afternoon and caught bass from each of them.

But the last spot was a doozy.

Dana had walked to one end of the pond; Herbie and I went toward the other. There was a willow tree growing out from the bank. Various stickups protruding from the surface indicated a veritable jungle of structure below. I was rigged with a seven-inch blue plastic worm. Herb was using one of his homemade top-water baits.

We cast almost simultaneously.

No sooner had Herb's lure begun sputtering along the surface than it disappeared into the maw of a big bass. At virtually the same instant, I felt the unmistakable *tap-tap-tap* of a bass down below and set the hook. We wrestled our prizes to the bank and laid them out, side by side.

Talk about twins! Both were in the five-pound range. His was maybe half an inch longer, but other than that, it looked as if they'd been stamped out with a cookie cutter. As we trudged back across the pasture to Herb's van, Dana mentioned the country store and suggested we go there to get a soft drink and weigh these babies. We covered them with ice in the cooler and drove away.

When we arrived at the store, Dana and Herb walked inside, Herb proudly toting his fish. I hung around in the gravel parking lot until they went through the door. Then I started grabbing rocks and pushing down the now-dead largemouth's throat. Once its belly was sufficiently distended, I marched into the store as well.

Herbie already had his fish on the scales. The needle showed five pounds, two ounces.

He removed his bass. I laid mine across the sheet of freezer paper the store owner had positioned across the scale seat. The needle flew over to five pounds, four ounces. Smiling broadly, I reached for my prize.

"Wait a minute!" said Herbie. "Somethin' ain't right here. Your fish is smaller than mine. How come it weighs more?"

He squeezed its belly and wrinkled his brow. Then, holding the fish by its tail, he shook it sharply.

Plink! Plink!

A couple of pebbles hit the wooden floor.

Herb shook the bass again.

Plink! Plink! Plink!

He picked up one of the rocks, rolled it in his fingers, then cut a mean glance toward me.

"All right, Mister Expert!" he demanded. "How do you explain these?"

I didn't bat an eye.

"They look like kidney stones to me," I said. "I reckon I did that poor ol' sick fish a favor by puttin' him out of his misery."

Then I grabbed my bass and headed out the front door without a backward glance.

I just hope Saint Peter thought that little trick was as funny as I did. Otherwise, my window shade is going to be sagging heavier than ever.

Getting
to the Point

I have just gotten off the telephone with Dr. William Bugg, a professor of physics at the University of Tennessee. He was explaining a fundamental law of matter and motion involving the penetration of one object into another.

It's a simple principle, the good prof said, of force divided by area.

Let's say you have two cylindrical objects, A and B, and wish to insert them, perpendicularly, into the surface of Object C. Both Objects A and B are of the same overall diameter, but the insertion end of Object A has been honed to a sharp point, whereas the insertion end of Object B is blunt.

Given equal amounts of pressure, Object A will penetrate faster and deeper than Object B. This is because the pressure is exceedingly great on the sharpened end of Object A. On Object B, the pressure is diluted because it is spread over a wider area. Thus, Object B will not penetrate as well.

This is an inarguable law of nature. It is just as applicable in the year 2000 A.D. as it was in 2000 B.C., and it will be just as applicable in the year 10,000 A.D.—assuming the world has not gone totally berserk from over-exposure to television commercials, brussels sprouts, computer games, fishing spies, and other forces of evil.

If you want a practical example of how this principle works, go to your workshop and cut a piece of seasoned oak two-by-four, twelve inches in length. Sharpen one end to a V-point. Leave the other end flat.

Next, take off your shoes and socks and drop the two-by-four lengthways, from waist height, upon your bare right foot. Do the flat end first.

In, oh, three or four hours, after the initial pain has subsided and you are able to walk without a noticeable limp, try it with the V-shaped end on your left foot—after you have telephoned 911 and placed the emergency crew on red-alert. That should answer any lingering questions you might have about matter and motion.

Herb was not a student of physics. To my knowledge, he never carried on a conversation about force divided by area with Dr. Bugg or any other academician. He could not have begun explaining this phenomenon scientifically if you held a cocked .38 to his temple and told him he had until the count of ten.

However, when it came to inserting Object S (as in steel) quickly and deeply into Object F (as in fish), Herb was a regular Einstein. From the get-go, he preached the vital importance of using sharp hooks. As far as he was concerned, this was the cardinal rule of fishing.

Yes, proper gear selection was important to him.

Ultra-light spinning outfits and delicate fly rods were fine for subduing ten-inch hatchery trout from a tiny mountain stream, he believed, but they had no place in the sweat-and-grit world of extracting bass from the brush. For this chore, a rod with plenty of backbone, matched with line heavy enough to yank small trees out by the root, was the obvious choice.

Yes, lure placement was important, too.

Herb's casting accuracy, particularly in the years before failing health cursed him with the shakes, was nothing shy of phenomenal. His opinion of casting was the polar opposite of horseshoes. Close didn't count one iota. In his way of thinking, each presentation either scored a bull's-eye or else it was a complete miss.

I have seen him make casts that all but defied gravity

and the laws of nature. He could sidearm a lure against the surface of the water with enough force to make it ricochet—which is no big deal, you are saying, because any fool who ever skipped a rock can do the same thing—but then guide the deflected missile like it was on remote control. If it needed to shoot under a dock or overhanging limb and then abruptly slice to the right, he could summon the correct amount of body English with a casual flick of his wrist.

Don't ask me how he did it. I just know he did. I watched it too many times. It was an act usually followed by a terse Herbesque dictum. Such as, "See? Now how hard can that be? And quit shakin' that bush over yonder! We'll go get your plug unhung in just a minute. I've got a bass to catch. *Herr-herr-herr-herr!*"

But the best tackle and the most realistic lures and varmint-rifle accuracy weren't worth a dadburn thing, he insisted, if the hooks weren't sharp. Needle sharp. So sharp, as country comedian Brother Dave Gardner used to say about knives, "the microbes would scream when you blew across the edge."

Herb had utter disdain for roughly 99 percent of the hooks that came on store-bought lures.

"There oughta be a law against people sellin' junk like this," he'd usually harrumph upon inspection of my tackle box.

"What's wrong with that stuff?" I'd say back. "Those are some of the hottest baits on the market."

"Yeah, if you're fishin' for suckers. But if you want to catch bass, the first thing you'd better do is sharpen those hooks. A bass has got a mouth like steel plate. There's a good chance you're gonna miss him, even though he climbs all over your plug. The least you can do is stack the odds in your favor."

"What do you know about odds? That sounds like gambling to me."

Herb would give me one of his patented How-Come-You-Turn-Every-Pearl-of-Wisdom-Against-Me looks. Sorta like Archie Bunker used to do when his meathead son-in-law would ask a question.

"Here I am, wastin' a perfectly good afternoon on you when I could be catchin' my supper. And what do I get? A smart aleck, that's what. If I had a file, I'd sharpen a couple of these fence posts you call hooks. Then maybe—just maybe— you could enjoy the feel of a bass on the end of your line instead of settin' the hook and sayin' 'rock-elephant' cause he got off."

It was a grand ruse, of course. Always was. Herb had a file in his tackle box. He never left home without one. He'd no more go fishing without something to hone the points of his hooks than a cop would hit the street without a sidearm. And usually with a bit of begging and pleading, I could talk him into putting an edge on my hooks, as well.

If you've ever watched one of those TV fishing shows, or read a how-to story in an outdoor magazine, you've surely seen the quick test for determining the sharpness of a hook. You're supposed to try hanging the lure across the surface of a fingernail.

Not with pressure, you understand. You're trying to inspect for sharpness, not inflict bodily harm. Just hold a thumb or finger sideways, position the point of a hook on the nail, and let go. If the bait slides off, the hook isn't sharp enough. If it sticks, you're in business.

Herb's hooks were every bit that sharp. Even more so. I wouldn't be surprised if, on extraordinarily humid days, they couldn't hang in the air by themselves.

Ah, but with all pun intended, there is a stickler in

this equation. Sharp hooks are like a double-edge sword. They will impale the nearest available flesh, caring not one whit if that flesh belongs to fish or fisherman. More than once, Herb and I left bloody fingerprints on the seats and sides of Murph (although never in the quantity I donated during the Day of Disasters on Little River). Usually it was nothing more than a prick, followed by an "ouch!" or a "dammit!" depending on the prickee, and then the fishing would resume. After a reminder about "rock-elephant," of course.

But one day, Herb did humbly find himself on the receiving end of a sharp hook—"end" being the operative word.

He was working in his shop that morning, fine-tuning a selection of top-water baits to take on our river float the next day, and made the crucial error of leaving one of the lures precariously close to the edge of the bench.

In point of fact, that was the second of several crucial errors. The first had occurred a few days earlier when Herb noticed that the horn on his van had quit working, and he put off driving it to the dealership for immediate service.

Another error had occurred on the same morning Herb was working on the lures. It was his decision to bush-hog some weeds behind his house. His tractor battery was dead, so he had hooked it up to the battery charger in his shop. This was the same battery charger that happened to sit quite close to his workbench.

But perhaps the most costly error of the day occurred when Herb leaned over to unsnap the cables from the battery charger. In so doing, he inadvertently snagged one of the treble hooks from the plug that had been left close to the edge of the workbench. Apparently Herb just barely brushed the bench, but that was enough. Unbeknown to him, the lure attached a free ride in the baggy part of his trousers, just below the cheeks of his butt.

No telling when he might have discovered this stowaway. Perhaps he might have felt it banging against his trousers. Perhaps the unattached treble hook might have emitted a rattle, giving some audible clue to its presence. Perhaps the lure might even have dropped off before inflicting any harm.

But nothing so innocent occurred.

Instead, the plug was still firmly attached to Herb's britches when he toted the tractor battery to his van. The tractor was stored in a shed in the back of Herb's property, so rather than carrying the heavy battery all the way through the back forty, he intended to drive it out there in the van.

All went well with his plan initially. Herb deposited the battery in the rear of the van. He walked toward the front. He swung the door open. He slid onto the seat.

It was then that the sharpness of the hooks became oh-so-painfully apparent.

One set of trebles dug into the fabric seat and held on like a terrier. The other dug into Herb's hiney with the same conviction.

"I was pinned," he told me the next day. "There ain't no other way to describe it. I couldn't move an inch. I had one leg outside the van and one leg in. Ever'time I drew a breath, those hooks dug in deeper. It was like big hunks of meat were bein' clawed outta my butt."

You think he was in a bind then?

Naaa. Things could get a lot worse. And they did.

You see, Herb's wife, Mary Jo, wasn't home. And even if she had been, there was no way to summon her because of the malfunctioning horn.

Fortunately, Herb's van had an automatic transmission. Thus, he was able to turn the key, start the engine, and proceed veeerrryyy slowly down his driveway in search of assistance, van door open, leg dragging.

Miraculously, Herb managed to putt-putt his van out of the driveway and across the road. It came to a slow stop in the front yard of his neighbor, Carl McDaniel. Once there, all he could do is wait. The horn wasn't working, remember.

"I guess I didn't sit there long," he told me. "Maybe four or five minutes. But it seemed like forever in the hot sun. Finally, Carl looked outside and saw my van sitting there and came out to see what was going on."

Let me tell you what a swell neighbor Carl McDaniel is. Not only did he take a set of wire cutters and free the plug from the van seat cover, he also then got up close and personal and freed Herb from the offending lure.

"It was the first time," Herb noticed, "that a big one didn't get away. *Herr-herr-herr-herr!*"

By and large, technology has freed today's bass anglers from having to sharpen their hooks, at least the ones that come out of the box. Manufacturers use laser beams to whittle the points down to needles. Several excellent brands are currently on the market, but it was a Japanese company, Gamakatsu, that came out with some of the earliest, sharpest models. Compared with conventional hooks, however, the ones from Gamakatsu were outrageously expensive.

"Those things are high as smoke," is the way Herb described the price.

In 1991, Herb thought of a way to help lessen the financial blow. That was the year I visited Japan. Matsushita Electronic Components Company, a Knoxville-based arm of Panasonic, asked me to accompany four high school students on a cultural exchange program. When Herb found out I was going to Japan, he instructed me to not come home unless I brought a selection of Gamakatsu hooks with me.

There was not a tremendous amount of time on my itinerary to peruse Japanese sporting goods stores. In fact, all I did one day was walk into a store that appeared to carry fishing equipment, use crude sign language to indicate I wanted hooks, and plunk down the appropriate amount of yen on the counter. Without so much as glancing at my purchases, I folded the Gamakatsu packets into a paper bag and stuffed them into my baggage.

I gave them to Herb when I returned to the United States eleven days later. He was on the telephone within hours.

"You know those hooks you got me?" he asked.

"Yeah," I replied, "how are they? Are they sharp enough?"

"Oh, yeah, they're sharp all right," he said. "But what I wanta know is how those fellers in Japan go about their fishin'."

"What do you mean?"

"The hooks ain't got no eyes on 'em!" he said. "How you supposed to tie 'em onto a line?"

I couldn't answer the question. Indeed, it wasn't until I drove to Herb's home a few days later and inspected the hooks closely that I realized what the problem was. Turns out I had purchased him eyeless snell hooks. They were intended to be wrapped onto a nylon leader before being used. I explained the snelling procedure to Herb, even showed him pictures of the process in fishing books.

"You mean to tell me they go to all that trouble just to tie on a hook?" he asked, incredulously.

"Yep."

"Well," he sniffed, "it ain't no wonder they lost the war."

Ol'
Redemption

It was one of those hot, humid, horrendous August afternoons, a day so sticky it seemed the entire world had been sprayed with honey mist. The kind of day that wilts the temper out of steel. The kind of day that starts steamy at dawn and grows worse by the hour. The kind of day to be ventured into only by mad dogs and Englishmen.

And Herb.

"Ain't this wonderful weather?" he beamed when we met at the Golf Course. "Makes a man glad to be alive. If it would just get a little hotter, things would be perfect."

It was perfectly clear to me that he'd been in the sun too long.

"If it gets any hotter, I don't believe I can stand it," I said.

"You could stand a few bass on the end of your line, couldn't you?"

"Sure."

"Well then, get out of that air-conditioned truck and let's get movin'. Ol' Redemption needs to be back in the water."

"Who is 'Ol' Redemption'?" I asked.

"I'll show you in a few minutes," Herb responded. "C'mon, son! Why do you dawdle so much? I'll swan, it'dud be dark before we started fishin' if you were in charge of things."

I was starting to believe the man was part reptile. The cooler the weather, the slower he moved. He owned an ancient, blue, London Fog topcoat that came out of the closet and stayed on his back, along with various other layers of insulation, from the first hint of autumn frost until spring had a firm grip on the land. But let the mercury keep climbing higher, higher, higher—to the point most mortals slowed to a stupor—and his motor only revved faster. Global warming was never a worry of Herb's. He thought it was an excellent idea.

We shoved the Hub Tub into the water, and Herb switched on the electric motor. Off we roared with all the blinding speed and apparent progress of a caterpillar crossing a football field.

"Let me bring Ol' Redemption out and let him have a look at the surroundings," said Herb. "It's been awhile since he's seen the Golf Course."

With that, he opened his tackle box and extracted an Arbogast Jitterbug. This is a wide-lipped surface lure that has been producing stringers of bass for legions of anglers ever since it was developed, quite by accident, in the early 1930s.

Despite the fact the Jitterbug is one of the most popular top-water plugs of all time, it began its life as a deep-running crankbait. I learned that little nugget of information years ago from Dick Kotis, then the president of Arbogast Bait Company in Akron, Ohio.

The lure had been designed by Fred Arbogast and one of his employees, Brook Oertel. They mounted a cupped lip on top. But during tests, the lure continued to ride on its side, then angle up to the surface. No matter how many adjustments they made, their new creation continued to misbehave.

In what has been long lost in the annals of angling history, Arbogast and Oertel eventually looked at each other and said something on the order of, "Duh! Why don't we listen to what this thing is trying to say to us?"

They repositioned the lip to the bottom, fine-tuned it, and introduced it as the wobbling, noise-making "Jitterbug." It sold by the freight car load.

"Mr. Arbogast didn't keep good records in the early years of his business," said Kotis, who joined the company in 1957, "but even by conservative estimates, we sold three-quarters of a million Jitterbugs until well into the 1980s."

That's three-quarters of a million *per year*, he clarified. Even though Kotis retired in 1994 and the Arbogast company was eventually purchased by Pradco Lures of Fort Smith, Arkansas, the Jitterbug is still being manufactured. And is still catching bass.

Probably because it is idiot-proof. If there ever was an autopilot lure, it's the Jitterbug. You throw it out and reel it in. Kotis told me that literally hundreds of anglers had approached him at tackle shows down through the years with tales about catching their first bass on a Jitterbug.

"That's because it was the only plug their daddies or uncles or big brothers would let them use," he said. "They knew the kid wouldn't hang up a Jitterbug on an underwater stump. So the kid would just keep casting it until a fish struck."

Technically, there are differences in the way bass anglers retrieve the Jitterbug. Some, like Kotis, prefer a stop-and-go style. Others let it sit motionless for long periods of time, merely jerking it along on a slow trip back to the boat.

Herb, however, was one of the disciples of steady movement. He kept it swimming—*"glub-glub-glub-glub-glub"*—across the tops of submerged rocks and brush, always tensed for a strike.

The Jitterbug Herb held in his hand was one I had seen him use before. It was black with yellow vertical stripes along the sides. Yet I had never heard him call it by any particular code name, especially "Ol' Redemption."

Assuming, correctly, that a soliloquy was about to be delivered, I settled into my side of the Hub Tub's seat for the duration.

"This bait," the preacher began, "has been born again. It has been given new life. It has been redeemed. That's why I call it Ol' Redemption."

I was more confused than ever.

"Happened a couple of months ago, right at the start of summer," he continued. "I was down here at the Golf Course by myself one—"

"How come I hadn't been invited?" I broke in.

Herb ignored the interruption completely.

"—day and accidentally left my tackle box on the bumper of the van after I'd put my boat in the back. I knew better than to set that box there. Matter of fact, I kept tellin' myself, 'Don't leave the box on the bumper! Don't leave the box on the bumper! Don't leave the box on the bumper!' But, of course, I drove right outta there with the box still on the bumper. I blame you for it. I've been hangin' around the likes of you too much. I reckon some of your hardhead is startin' to rub off.

"Anyhow, I got home and started unloadin' my gear, and that's when I realized what had happened. I jumped right back in the van and drove back down here to the Golf Course. I couldn't have been gone more'n thirty or forty minutes. I figured the box musta rolled off when I pulled up onto the highway. But when I got back down here, it was gone.

"I parked and got out and looked all through the weeds and the ditch. Nothin'. In fact, I walked up and down the road a good hundred yards. Both sides. That box was not to be found. I figured I'd never see it again."

The painful remembrance brought a wince to Herb's tanned face.

"You ain't got no idea how awful I felt," he sighed.

"Sure I do," I said. "I've lost things before."

"But you've never lost anything as precious as this here Jitterbug. It 'bout made me sick. I'd caught so many bass on it, and we'd become such good friends, I couldn't hardly bear th'thought of somebody else ownin' it. Why, I stayed up nights worryin' about this poor ol' bait."

As I knew from watching him cast seventy-times-seven, Herb does not give up easily on anything. He assigned his gray cells to the task of locating his AWOL fishing tackle.

"If a fisherman found that stuff, I knew it was useless," he continued with his story, "but I kept hopin' it wasn't a fisherman who found it. Turned out I was right."

"Did you have your name and phone number inside the box, and the fellow who found it gave you a call?" I asked.

Silly me.

"Of course not!" Herb snapped back. "I didn't have my name inside the box. Wouldn't a'been no use. In the first place, I didn't want nobody knowin' what I was fishin' with. In the second place, I knew if anybody realized these were my baits, they'd not return 'em for sure. You know how people are always tryin' to find out my secrets."

"So what happened?"

"If you'd hush for a minute and let me finish, I'll tell you," he said. "Why are you always so impatient? You know, you'd catch a lot more fish if you'd just slow down and let things happen natural-like. But nooo! You gotta stay wound up like a seven-day clock all the time. It's a wonder you ever—"

"Will you get on with the damn story!" I shouted in desperation.

"Rock-elephant," he countered. "Just say, 'rock-elephant.' It'll calm you down. What would you do if you didn't have me around to teach you things?"

It was hopeless. I rolled my eyes and folded my arms and told him to continue.

"It came to my mind that the best place for somebody to sell those baits would be at a flea market or garage sale. So ever' weekend, I started drivin' around and checkin' 'em out. Took me about a month, but I finally hit pay dirt. I was walkin' down the aisle over at Green Acres flea market, and there it was."

"Your tackle box?"

"No, Ol' Redemption! But it wasn't redeemed just yet. I asked the ol' boy what he'd take for that plug, and he said three dollars."

"You didn't pay that much, did you?"

"Well, I sure would have," Herb said. "I'dud probably paid ten to get that plug back. It runs as good as any Jitterbug I've ever owned. But I did haggle with him a little and got him down to two-fifty.

"Soon as I got my hands on Ol' Redemption—that's when I named him—I asked the ol' boy, 'You got any more plugs like these?' And he says, 'Sure, I got a whole box full of 'em.' And out he came with my tackle box."

"You reckon he's the guy who found it along the road that day?" I queried.

"I don't know," Herb answered. "I didn't want to ask. All I wanted t'do was get my hands back on that box and those plugs. I wound up payin' over twenty dollars for the whole she-bang, but it was money well spent."

Mercifully, the sermon was over. It's a good thing, too, for we had just arrived at the mouth of the bayou. For all I know, Herb planned it that way. In any event, he announced it was time to officially send Ol' Redemption back into the fray. He held the plug at eye level and spoke to it so seriously, I could have sworn the thing lived and breathed and had a conscience.

"Now you lissen to me," Herb instructed. "I have saved your life. You have been redeemed. Don't go out there and do anythin' foolish like getting' hung up on a limb. You go out there and catch me a bass."

Slowly, methodically, Herb and Ol' Redemption worked the first brushy point of the bayou. Nothing.

They explored the back point. Nothing.

They investigated the rock pile just beneath the surface, on the left edge of the creek channel. More of the same nothingness.

A worried look came over Herb's face.

"I sure hope Ol' Redemption didn't take on any bad habits while he was gone," he pondered aloud. "This could be more serious than I expected. I might have t'do some real prayin' over this plug to get it back on the straight and narrow."

Not quite. At the big brush pile at the fork of the island, Ol' Redemption got back into synch. Herb cast it behind, and to the right, of the thickest tangle. Soon as Ol' Redemption splashed down, Herb began reeling steadily.

It started back, wallowing in a tight pattern from side to side.

Glub-glub-glub-glub.

"Ain't that pretty music!" Herb exclaimed. "I'll swan that's the best he's ever sounded."

"Maybe Ol' Redemption had you fooled," I offered. "Maybe he wasn't off learnin' bad habits at all. Maybe he was doin' graduate studies in glubology. Y'know, perfectin' his pitch and tone."

Right then, a giant swirl enveloped the Jitterbug. Herb set the hooks instinctively, and a good three pounds of largemouth rocketed into the air.

"Yeeee-haw!" Herbie screamed. "Ol' Redemption is back!"

I was never happier in my role as net man. One scoop and the bass came aboard. The poor thing never had a chance. Ol' Redemption was crossways in its mouth. Two hooks each from both trebles were securely attached. The only way that bass could have escaped was by breaking Herb's ski rope, and that wasn't about to happen.

Herb was euphoric with glee. He burst into a heartfelt chorus of the country music song, "My Baby Thinks He's a Train," a tune which, for reasons only known to him, was reserved for special moments of joy. Holding the fish between his knees, he used pliers to back out the hooks.

When he wasn't singing a verse of "My Baby Thinks He's a Train," he would whistle it, loudly and emphatically, in what I called "wood-chopping whistles." Every time he paused long enough for more air, the first blast out of his lips came out in a *whoosh!*—just like an ax or splitting maul whipping through the air.

Normally, I would have used Herb's down time to my advantage. With him distracted by the chore of hook removal and engaged in joyous singing, I could get in at least two or three casts before he'd start swinging again.

But not this time.

Herb was just starting to grow weak from his thrice-weekly dialysis sessions, and I was slowly coming to the dark realization that way out yonder, somewhere down the path, our camaraderie would eventually come to an end. So I sat there in the Hub Tub, soaking in all the sights and sounds, trying to find a way to preserve this moment forever.

And it dawned on me that Herb was right. Muggy as it may have been, this was indeed wonderful weather. Those poor souls back in their air-conditioned cars and offices didn't know what they were missing.

Maybe all they needed was a little redemption.

Horse
Trading 101

A wild turkey's ability to hear is nothing shy of phenomenal. At one hundred yards, through dense forest, a gobbler can detect the faint "cutt," "cluck," or "yelp" of a hen—or a hunter attempting to make the same sounds—and determine the location with a precision that is all but surreal.

When the amorous he-turkey picks up the siren's sound and starts walking or strutting toward it, he is not heading "over yonder." That is because he knows the signal is not merely coming from "over yonder."

He knows it is not coming from "just below the crest of the ridge, over by the moss-covered rock," either. It is much more specific than that.

He knows it is coming from "the base of the white oak, *Quercus alba*, eighty-two inches in circumference, located nineteen feet northwest of the moss-covered limestone rock that juts out of the ground at a twenty-six degree angle, thirty-seven yards below the highest peak on Baker's Ridge." Give or take a foot.

The turkey does not understand his mission in those exact words. He doesn't need to. This is his home turf, and he has been intimately acquainted with every square inch of it, on his own terms, since he pecked out of the shell. From the instant the sound wave of the she-turkey's call strikes his ear drum, he picks it up, filters out the solo from a nearby wood thrush, and locks on with the accuracy of a sniper's rifle.

If, at any point en route to the site, that same feathered radar detects a quiet cough, the scrape of a camouflage coat against tree bark, or the sharp *click* of a shotgun safety being snapped off by some imbecile who doesn't know how to stifle the noise with his fingertips, the turkey will not conduct a debate with himself about what he did, or perhaps did not, hear. He will simply turn around and walk away, no questions asked.

You have to be on the receiving end of one of these damning phenomena—and I have—to fully appreciate it.

Incredibly, however, it is possible for one human being, fully camouflaged, to whisper into the ear of another, similarly dressed and sitting shoulder-to-shoulder, as the turkey closes the distance. I have witnessed this phenomenon, too, and have either launched, or commanded to be launched, a load of No. 4 shot to bring the show to a sudden and violent conclusion.

Realistically, calling this communication a "whisper" takes liberties with the King's English. A "whisper" is what occurs in theaters and opera halls and is usually followed by a tart "shhh!" from three rows away. In the context of turkey hunting, a "whisper" is nothing more than an audible thought, transmitted from one party to the other in the most muted of tones.

All of which is a terribly long explanation of how, and why, I was mouthing, "He's about to make us, Patrick! You'd better shoot! Now!"

I wasn't dealing with a rookie. Patrick Hubbard, Herb's oldest grandson, had been turkey hunting with me for three years and had bagged a bird every season. He knew the drill—don't move, don't blink, don't so much as even breathe—as well as any student I'd ever tutored. Just as the gobbler extended its white-domed periscope an inch or two higher to inspect the lumps at the base of the tree twenty yards away, Patrick's twelve-gauge magnum roared.

He and I leaped up excitedly and sprinted toward the fallen bird, arriving scant seconds after the pellets themselves.

This was a splendid gobbler. At the country store on our way home, he pulled the scales to just over nineteen pounds. His beard taped a full ten inches. He had behaved in textbook fashion, too—a classic 9 A.M., starved-for-female-companionship king of the woods who answered my initial call with a loud gobble, and then sounded off, virtually on command, every five or six minutes as he worked our way.

Patrick and I had flexed a lot of boot leather that April, and now it was time to celebrate. As soon as we hiked off the Rhea County mountain and fired up the truck, he reached for his cell phone and declared, "Let's call Papaw and tell him all about it."

Patrick held the phone away from his ear while he drove, winking and grinning as his grandfather sermonized on the other end of the line. Over on my side of the vehicle, the fulmination came in loud and clear.

"Well, it's about time the two of you did some good! I'll swan, I don't know why you waste all that time chasin' after a turkey when you could go to the grocery store and buy one that's already plucked and cleaned."

"You could go to the store and buy fish and save yourself a lot of effort on the lake," Patrick countered.

"That's different!" Herb snapped back. "It's bad enough that ol' Fingernail fools around with feathers when he could be catchin' bass with me. But now he's got *you* hooked, too!"

"We're gonna clean it over at Sam's," Patrick said when his grandfather paused for air. "You wanta come over and see it?"

"I don't know if I have time," Herb's voice crackled through the receiver. "I've got so much goin' on right now. You'ins'll take an hour to clean that thing, and all you'll talk about is feathers. Have y'all forgot how to fish?"

"Papaw, we can fish after turkey season has closed," Patrick said.

"Not if you keep hangin' around with Fingernail! He'll be the ruination of you. But I reckon since you'ins have interrupted my peace and quiet already, I might as well blow the rest of the mornin'. I'll be over in a little while. Bye."

"That means you better step on it," I said to Patrick. "I'll guarantee the old coot's halfway down his driveway right now."

Amazingly, we beat Herb to my house by a good ten minutes. I started rounding up fillet knives, shears, plastic bags, and other items we'd need for the bird-cleaning process. Just as I walked past the workbench in my garage, I spotted two old magazines.

I knew those magazines well.

They were *Sports Afield*, circa 1939, and had come from Herb's collection almost a year earlier. Each contained a duck-hunting story he thought I'd enjoy reading. He had handed them to me—strictly on a loaner basis, of course—on one of our fishing trips the summer before.

"Do *not* let Herbie leave here without takin' these magazines," I told Patrick. "I've been meanin' to give 'em back to him for months, but they always keep slippin' my mind. He's probably forgotten all about 'em by now, too. But I simply *must* give them back."

Herb showed up about that time, and the three of us fell into the standard post-hunt photography session. I snapped pictures of the turkey. And Patrick holding the turkey. And Patrick holding the turkey and his shotgun. And Herb and Patrick holding the turkey. And Herb and Patrick shaking hands over the turkey. I even dug out my tripod and, using timed shutter release, took pictures of Patrick, Herb and myself holding the turkey. Whoever processed the film surely had to wonder what was so special about a dead bird that we would exhaust a twenty-four-exposure roll on the same subject.

Once the camera was stowed, Patrick and I began concentrating on the bird.

Herb began concentrating on my garage. Specifically, the shelf holding my fishing tackle.

Why? Because it was there.

Even though Herb had repeatedly pawed through every tackle box I owned, he simply *had* to poke around some more. Something might have escaped his eye on earlier perusals, something he'd try to talk me out of.

I was up to my elbows in the body cavity of Patrick's turkey when Herb walked into the back yard. He was carrying a graphite spinning rod and a Cardinal-V reel that had been sitting in my rod rack for so long, they were coated in dust.

"What'llya take for this junk?" he asked.

I hardly glanced up from my chore.

"It's not for sale. I use that rig all the time."

"You do no such thing!" he snapped. "It's been at least five years since you carried this outfit. Why, I could plant corn in the dust and dirt that's built up on it—and grow a good crop, too. You don't need it, so why don't you let me take it off your hands?"

Herb was correct. To a degree.

Indeed, it had been years since I touched that rig. The line on the spool was probably dry-rotted by now. I didn't use that rod and reel. I didn't need that rod and reel. I could not have cared less about that rod and reel if they dropped off the face of the earth.

But I knew one thing for certain about that rod and reel: Herb wanted them. This golden nugget of information gave me immense bargaining power.

I am not the world's greatest poker player. But I have sat around enough card tables—and done enough horse trading—to know how to camouflage my excitement upon the receipt of a sudden windfall. In this case, the dealer had just handed me a straight flush, and the pot was mounded in chips.

"I'll give you fifty dollars," Herb offered. "It ain't worth half that, but I reckon I owe you somethin' for callin' in that turkey for Patrick—even if you have 'bout ruined him with feathers."

"Fifty, my ass!" I sneered, still concentrating on the bird. "I'm tellin' you, Herb—the outfit ain't for sale. I need it."

"Sixty," said Herb.

I can remember thinking there are so few times in one's life that the cards fall this way. Oh, to have my performance captured on video, the better to savor it on some cold winter day in years to come! Clint Eastwood couldn't do better at his "Dirty Harry" best!

Carefully, I laid the fillet knife in the grass. Methodically, I began wiping gobbler goo off my hands. I didn't utter a word for a few seconds.

"That's one of my best outfits," I finally spoke in a calm, even voice. "But if you want it so badly, you can have it for a hundred and a quarter."

"*What?!*" he thundered. "Why, that's robbery!"

"Then put it back in the rod rack."

Herb retreated to the garage at a dead run.

"He'll be back out here in about three minutes," I said to Patrick. "I've got him gut-hooked. All he's doin' right now is runnin' with the line."

I held the turkey while Patrick began de-winging with shears. In almost no time, Herb was standing over us again. Just as I had anticipated, he was holding the rod and reel.

"Seventy-five, and not a penny more," he said.

I turned and looked him straight in the eye.

"Herbie, the price is 125, firm. If you want it, fine. If not, put it back in the rod rack."

Off he went again, fuming louder than before. I winked at Patrick, relishing every delicious second of the conquest. We finished dressing the turkey, dropped the cleaned carcass into a plastic garbage bag, and were filling it with ice when Herb reappeared. With the rod and reel, of course.

"One hundred and that's my final offer," he said. "Take it or leave it."

"I'll leave it."

I don't believe I could have kicked Herb in the shins and made him any madder. He stewed. He fussed. If Herb had been given to strong language, he surely would have gone far beyond "rock-elephant." But I had him where the hair was short, and he knew it.

"One hundred, and I'll throw in two Dartin' Rs," he blurted.

Wow. Now I knew I *really* had him.

The Dartin' R was a fabulous crankbait made by the late Ron Ripberger, a Kentuckian. Herb coveted his Dartin' Rs like they were made out of platinum. Sometimes he'd give me one for Christmas or on a birthday, but always only one. For him to offer two, right off the bat, was a sure sign of surrender. He was warm putty, waiting to be molded in my hands.

"One hundred dollars," I said, casually rolling a toothpick in my teeth, "and *seven* Dartin' Rs."

"*Seven?!*" he hollered. "Are you are out your mind?"

"Nope. That's the price."

"Well, I ain't buyin' at that price!" he snapped.

And back into the rod rack it went. With a thud this time.

Oops, I thought. Bad move on my part. I must have upped the ante too quickly. Now what to do?

And that, gentle reader, is the exact moment my eyes fell on those two copies of *Sports Afield*. The same two I had been meaning to return to Herb for all those months.

"One hundred dollars and seven Dartin' Rs," I said— "and I'll throw in these two magazines for boot."

Herb didn't pause to consider the bid.

"Deal!" he said, snatching the magazines from my grasp. "You're a robber and you know it, but I'll take it anyway."

He dug five twenties out of his wallet, withdrew the rod and reel from the rack, and headed toward his truck.

"Meet me at the Cherry Pit for lunch on Friday," he said over his shoulder. "I'll bring your dang ol' Dartin' Rs then."

He nearly burned rubber peeling out, and I nearly exploded trying to hold in my laughter until he was down the road.

"Finally!" I shrieked to Patrick. "For the first time in history, I whupped your Papaw in a deal! It was crooked as a dog's hind leg, but by golly, I got him! Yee-haw!"

Patrick couldn't help but laugh, too. "Wonder how long it'll take him to discover that you hoo-dooed him?" he asked.

"We'll know on Friday."

Herb was uncharacteristically late that day. Patrick and I were already seated at the Cherry Pit when he trudged in. He was carrying a small paper sack. There was a scowl on his face as deep as the Grand Canyon.

"Patrick," he began, "have you ever heard of a man so lowdown and so rotten, he'd trade your own stuff back to you?"

I broke into laugher. "When'd you discover it?"

"When I went to put those magazines back into my collection," Herb answered. "I thought they looked familiar. Then I realized those were the same ones I let you borrow." He turned to his grandson.

"What kind of a feller would do that to his friend, huh? Patrick, are you takin' all this in? Do you see what type of sorry character you've been runnin' with?"

"Forget Patrick," I said. "Where are my Dartin' Rs?" Herb's face went blank.

"You ain't expectin' me to honor the deal after you cheated me like that are you?" he said.

"I most certainly am. A deal's a deal. Give."

Grudgingly, Herb shoved the paper sack across the table. I picked it up and opened it slowly, fully expecting to find contents that I richly deserved. Like maybe a vial of rat poison or a rattlesnake.

Instead, there were Dartin' Rs. Seven of them. I counted.

"Even if I get hoodwinked and tricked by a scoundrel, Fingernail is right," Herb said to Patrick. "A deal's a deal. Now, let's eat. I'm hungry."

I cannot begin to tell you how badly my conscience hurt during that meal. As much as Herb and I teased and played tricks on each other, I realized I had carried this one too far. Here he was, my best friend in the world, and I had—with full malice aforethought—cheated him. Yet he had managed to blow it off. Already, Herb was chattering about our next trip. My conscience was thrashing me so badly, I almost gave him those seven lures back.

Almost, I reiterate.

Way down in my psyche, deeper than the farthest reaches of my conscience could plumb, a still small voice was speaking—in roughly the same muted tones one turkey hunter uses to communicate with another. When Herb and

I met at the Golf Course later that week, I finally heard what that voice was saying.

I tied on the first Dartin' R out of the bag, cast it toward the bank, and started retrieving. Instead of biting into the water and digging deeply, the plug rolled onto its side and cartwheeled back to the boat.

No big deal, I thought to myself. Many homemade baits require fine-tuning. So I reached for my pliers, adjusted the eye slightly, and cast again. With the same results.

Herb appeared not to notice any of this.

I changed lures. Same thing.

The awful truth washed over me like a spill of toxic waste. There was no sense trying the other five baits.

"None of these Dartin' Rs will run straight, will they?" I said.

"Huh?" he responded, innocent as an angel. "You talkin' to me?"

"Every one of these things runs crooked, don't they?"

"Oh, *those!*" he beamed. "They're supposed to run crooked. They're designed that way, you know, for runnin' around stumps and boat dock posts. *Herr-herr-herr-herr!*"

I busted out laughing, too. Even harder than I laughed when Herb drove off with his magazines.

"You no-good, lyin', cheatin' sonuvabitch!" I yelled in mock anger.

"Rock-elephant," said Herb. "Always remember to say 'rock-elephant' when you're angry."

He made a cast. Then he turned to me, grinning like the proverbial mule eating briars.

"There's not hardly ever a Dartin' R that runs bad, but I've gotten a few over the years," he began. "I've been tryin' to get shed of those sideways-runnin' things forever! You and the Good Lord finally showed me the way. Ain't it wonderful? I got rid of some junk, and you got taught a good lesson. *Herr-herr-herr-herr!*"

You know that old saw about not playing cards with anybody named Doc and not eating at a joint named Mom's? Let me add one more.

Don't horse-trade with preachers. They've got too many connections in high places.

Aromatherapy is a New Age form of healing that uses familiar, pleasant scents to carry the patient back to a happy, less stressful time in his or her life. The sweet smell of honeysuckle in bloom might do the trick for one person. Grandmaw's fried chicken might work for another. Ditto a splash of Old Spice aftershave or White Shoulders perfume.

But there's an adverse side effect of this treatment. What smells wonderful in some quarters might be perfectly repulsive in another.

Lavender, for instance.

For some folks, this is a calming, soothing scent, reminiscent of pleasant walks through the garden on a sun-splashed June morning. Lavender—yuck!—reminds me of funerals. Even a whiff of it can send me into depression. I want no part of it.

So consider yourself warned for what comes next: I happen to enjoy the dank, mossy smell of barren mudflats, freshly exposed to the air.

This is not necessarily a pleasant aroma. Certainly not in the manner that, say, baking bread pleases the senses. I have never sniffed a nose full of mudflat air and said, "Yummy! When do we eat?"

Nonetheless, whenever I walk—or slog, as the case may be—across a barren flat of this dark, oozing goo, my quality-of-life meter skips into overtime. This is surely due to the fact that some of the most fun moments of my outdoor life have been spent on mudflats, thanks to the Tennessee Valley Authority.

In my region of southern Appalachia, God didn't create mudflats. The government did. Starting in 1933, TVA began to harness running water throughout the Tennessee River Valley. This fifty-year-long feat was accomplished by a series of dams that turned 652 miles of the main stream, plus 148 miles of tributaries, into one giant, serpentine pool with 11,000 miles of shoreline.

As a result, innumerable floods were prevented—assuming, of course, one buys into the federal concept of intentionally and permanently inundating a river bottom to prevent it from being submerged periodically by Mother Nature. Commerce and industry in a seven-state region were jump-started—assuming, of course, one also factors in a few billion bucks worth of government stimulation. And from an outdoor enthusiast's point of view, this infusion of concrete and steel created an aquatic playground, the likes of which had not been visited upon the earth since Noah was counting animals two-by-two—assuming, once again, one discounts the loss of free-flowing water recreation and concentrates solely on man-made environments.

Conventional recreationists enjoy TVA's lakes during warm weather. They fish, they ski, they camp, they swim, they drive powerboats. I can relate to all of the above. These are noble activities. I recommend them highly.

But it is after Labor Day, when 99 percent of the South has turned its recreational attention to football, that we mud-sloggers truly come into our own. This is roughly the time of year TVA begins pulling its reservoirs in anticipation of mid-winter floods.

Depending on the particular lake and river system, drawdown is either a slow affair that takes the better part of four to five months to complete; or else—clunk!—it all but occurs overnight. Depending on one's definition of aesthetics, what's left is either a dirty bathtub ring or a fragrant playground.

Newly retired Yankees who purchased their property in the midst of summer, when gentle waves lapped at grassy lawns, almost universally belong to the former group. Duck hunters belong to the latter.

I am neither retired nor of Yankee extraction. I do own approximately ten dozen duck decoys and five pairs of hip boots. That should give you a hint about my definition of aesthetics.

Hunting at the edge of these barren wastelands is not a gentleman's sport. It is a nasty, muddy undertaking. By the mid-point of any given waterfowling season, every duck boat in the valley will be coated, inside and out, with red clay and river-bottom gumbo.

Since the nearest vegetation is usually hundreds of yards away from the waterline, there is no readily available brush to build a traditional blind. Hauling limbs and trees to the water is a waste of time for two reasons.

In the first place, TVA will raise or lower the lake on a day's notice, submerging your architectural gem or stranding it a gunshot-and-a-half away from your decoys.

In the second place, a pile of freshly cut cedars in the midst of a mudflat does not fool a duck in the least, no matter how good your calling and how convincing your decoy spread. It stands out like a Republican at a Truman Day rally. You might as well erect a flashing neon sign that reads, "Warning! Duck hunter hidden nearby!" That is why mud-flat hunters either sit or lie in the mud and cover themselves with burlap and canvas.

Do this for fifteen or twenty winters—especially in bitterly cold Januarys when the mallards and black ducks pour into the valley like locusts—and you will develop a keen appreciation for the smell of burnt gunpowder, thermos coffee, wet Labrador hair, and mud.

But Herb showed me how to appreciate mudflats long before the first hint of frost. Indeed, fall had not officially begun when he called the house one September day and announced we would be fishing the "Slop Hole" the following morning.

"The what?" I asked. "Where in the world is that?"

"You amaze me," he replied. "How can a man claim to know something about catchin' bass in this part of the country and not know about slop holes?"

"I've never heard of them. Where do we launch the boat?"

"We don't," said Herb. "We go in by foot. Be prepared to get muddy."

"Where is it?"

"You let me worry about that. Just wear old clothes."

That's as much clarification as I am ever going to reveal publicly, too. Slop holes are located on virtually every TVA lake. Just look for old creek channels as the lake recedes. Some are more easily accessed than others. Some provide better fishing than others. I was lucky that Herb showed me his best ones. If you discover them on your own, fine. But I owe it to his memory to keep the precise locations a secret.

Correction; the previous sentence is a bald-faced lie. Herb's memory doesn't care one whit if everybody in the Tennessee Valley starts fishing his own personal slop holes. But I care. I still fish them faithfully. They are just as productive as ever, and I intend to keep them that way. We slop hole purists have feet of clay—feet that are covered with clay, come to think of it.

Herb had chanced upon his maiden slop hole years earlier while making a sewing machine delivery. Driving along a country road that paralleled a reservoir, he pulled over to the shoulder and stared out at the flat, muddy vastness, cloven only by the winding channel of an old creek.

Water still bubbled down the original trough of the stream, like it had for the eons before Congress chartered TVA. And just where the flow entered the reservoir, Herb saw a sight that made his casting arm twitch. Dozens of predatory fish had herded schools of shad minnows into the shallows and were feeding ravenously.

Herb suspected these predators were bass. When he came back the next day, equipped with old tennis shoes, his casting rod, and a selection of small crankbaits, his hunch proved correct. Naturally, he kept the secret from me until he felt I was responsible enough to take a blood oath of silence.

"Now don't go writin' any stories about these places," he warned. "Not for the newspaper or magazines or nothin'. These are our own spots. You've ruined every decent place on the lake already."

"How could I have done that?" I wanted to know. "I never give out an exact location in any of my stories."

"People still find out," he said.

"Well then, what's to keep them from finding out about slop holes?"

"They've gotta work too hard for their fish. It's a lot easier to stay in a boat."

No truer words were ever spoken. Bass fishing in slop holes is somewhat akin to wrestling greased pigs. At the end of the game, it's rather difficult to separate winners from losers. In the case of slop hole angling, however, the winners usually have a stringer of bass to clean before they hop into the shower.

Aside from the fact you often are in mud to your shins—and may have to cross half a mile of territory to find the best action—this is an incredibly simple exercise. Imagine casting into a drainage ditch. Any small minnow-like lure will suffice. Herb and I used Gay Blades, Spots, and Rattle Traps most often. I found it easier to carry everything in a fly-fishing vest rather than fool with resting a tackle box in the soup.

Lure color rarely matters because the water in these situations is heavily stained or chocolate-milk muddy. No doubt the bass zero in on vibrations from the lure as it swims through the water. They strike like a bolt of lightning.

Most of them are smallish—two pounds or less. But every now and then, even the shallowest of slop holes will cough up a leviathan of five or six pounds. Makes you wonder how they can even stay hidden in such sparse cover, let alone actively feed.

Herb and I began fishing slop holes more and more often in our later years together, even as he grew physically weaker. Especially after a session of dialysis, it might take him five minutes to shuffle from his car to the inside of a restaurant. But as soon as we arrived at a slop hole, he would bound out of the vehicle vibrant and refreshed. You can't convince me there's not a healing quality in that mud.

Sometimes, though, Herb bit off more than he could chew—of slop holes, I hasten to clarify; the man never met a platter of fried chicken he couldn't annihilate. We were at the "Lower Slop Hole" one afternoon, catching bass virtually at will, when I heard his cry for help.

I slogged and slid fifty yards up the creek bottom to where he stood, mired in partway up his shins. He had worn galoshes that day, thinking he would get better traction than in his tennis shoes. I, on the other hand, had opted for ankle-tight hip boots. Those are still my favorite for wading in the syrupy mess.

"Fingernail, I can't budge!" he exclaimed. "Help me get outta here!"

"What's it worth to you?" I asked.

He cut me one of "those" looks—like fathers give when their children talk back or push the curfew envelope too far.

"I ain't teasin'," he said. "I really need help."

"I ain't teasin', either," I lied. "If I've got to stop my fishin' and rescue somebody who ain't got sense enough to know not to wear galoshes into a mudflat, the least he could do is buy me supper."

"Supper, lunch, breakfast, whatever it takes. Just give me a hand."

I waded to one side of Herb and surveyed the situation. Sure enough, he was stuck like glue. I reckon if he had been there alone, he would have stayed until the buzzards picked his bones.

I tried tugging on his good arm—the one without the shunt, the one not riddled with needle marks from dialysis. All that did was nearly dislocate it from the socket.

"I'm gonna have to pop you out like a wine cork," I said, moving behind his back. "Spread both arms."

He did. I enveloped him in a bear hug at the shoulders and leaned back. Slowly, he began to move.

I leaned harder, pressing down on the back of my legs with all my strength. With a slurping *"plooooop!"* he rocketed free. Naturally, we both fell back into the mud.

By the time we both managed to squirm around and stagger to our feet, we were caked in slime and silt. I've never been dirtier in my life. Neither, I assume, had Herb. I don't know which of us was laughing louder. We were like two school boys during recess on a muddy playground.

I got dirty a few weeks later, but there wasn't anything funny about it. At least not at first. Herb had dialyzed that morning and was feeling particularly weak by the afternoon. We decided to go fishing anyway.

"You go on down yonder," he motioned with a sweep of his rod when we arrived at the "Upper Slop Hole." "I'm gonna stay here, closer to your truck."

"You sure you're gonna be all right?" I asked. "Once I get down there, I'll be out of sight."

"No, I'm fine," he replied. "You go on."

I took off at a dead slop, stopping every hundred yards or so to look back and check his progress. Shortly before I rounded the bend, I saw him wrestle a lively bass to shore. He's OK, I said to myself. That fish'll do him good.

I didn't stay long on the lower end. Perhaps not more than thirty or forty-five minutes. The fishing was decent—I was catching and releasing foot-long largemouths on every tenth or twelfth cast—but something told me I had better stick closer to Herb, just in case. When I trudged back around the bend and looked upstream, he was nowhere to be found.

"Herb!" I yelled. "You all right?"

The echoes from my shouts faded away into silence. Immediately, my heart flew into my throat. We instant panickers can do that on a second's notice. Oh, Lord! I thought. I've gone off and left him, and he's had a heart attack or drowned or something!

I took off running, which is a championship stretch of the word. My hip-booted feet were moving quite rapidly, but there was little forward motion. It was more like those cartoon characters who churn up mounds of dust before they race away. Except there was no dust in this case. All I generated were huge globs of mud that flew in every direction as I splattered back up the edge of the creek channel.

"Herb!" I yelled between jerking breaths. "Where are you?"

My truck was still parked at the crest of the hill, but Herb had absolutely disappeared from the pancake-flat landscape below. My stars! Had the earth opened and he been swallowed up?

It took another ten minutes of plowing and slogging, but I finally reached the base of the hill. I began clamoring across the rocks, taking two and three per step. Not until I reached the top, still expecting to find him prostrate on the ground, did the sound of his snoring fill my ears.

Herb did not snore like most old men. He bellowed.

Even now, I can't believe I hadn't heard the roaring from down below. Adrenaline will do that to you.

He was in the bed of my truck, curled up like a baby. Three bass, tethered to a rope stringer, lay beside him. The largest was a good four pounds. I banged my fist hard on the quarter panel, and he sputtered awake.

"Now why do you want t'ruin a man's rest like that?" he yawned. "That was the best sleep I'd had in weeks. Where's your fish?"

"Th'hell with fish!" I shouted. "You sumbitch, you scared the crap outta me! I thought you had drowned!"

"Oh, you thought nothin' of the kind," he poo-hooed. "You were just jealous, Fingernail, 'cause I was doin' all the catchin'."

He held up the largest bass for my inspection.

"I want you to look at this big'un right here. It's a shame you couldn't have been here with me when I caught him. I tried to get you to stay and fish here with me, but noooo. You had to go off galavantin' down there where the bass are little."

He looked me up and down.

"I'd suggest we drive into town and get somethin' to eat. If I ain't mistaken, I still owe you a meal for that little rescue job the other day. But I can't go nowhere with the likes of you now. Look at you—all muddy and sweaty. I'll swan if you ain't a mess! Shew! Why, you smell worse than this slop hole. *Herr-herr-herr-herr!*"

I didn't know whether to kick his butt or hug his neck.

That was a long time, and a lot of bass, ago. I think about it every time I drive past the spot or get a whiff of mudflat cologne. And sometimes even now, when I'm down in a slop hole and the wind's blowing right, I swear I can hear Herb snoring.

I hope he has pleasant dreams.

Mesmerizing

W hen I came banging and bouncing into the launch ramp at Friendly Hollow with Murph lashed in the bed of my truck, Herb was waiting. As usual.

He gave me a thorough what-for because I was ten minutes late. As usual.

He was wearing a camouflage hat, a camouflage T-shirt, and long dark pants, and loudly complained that my light-colored shirt and shorts—he called them "bikinis"—would spook every bass in a three-county region. As usual.

He already had his casting outfit, his cigar box assortment of surface lures, his seat cushion, and his water bottle laid out beside the ramp. As usual.

But then there was something most unusual. Resting alongside his gear was a black, telescoping, fiberglass pole about as big around as a walking stick.

"What's that thing for?" I inquired.

"It's for you not to worry about," he answered. "Let's launch this boat and get outta here before people start noseyin' around and askin' questions. They're even worse than you about pokin' in things that ain't none of their business. It just so happens that I intend to fish with this pole this afternoon."

"What with—redworms and a cork bobber for bluegills?"

Herb paid no attention to my remark. Instead, he began untying the rope that held Murph in the back of the truck. "It looks like I'm gonna be fishin' alone. I'll swan, it takes a long time for you to get movin'."

We slid the johnboat into the water, clamped an electric motor to the transom, and loaded the gear. When I returned from parking the truck, Herb was seated in Pole Position.

Huh? Herb in the front of the boat? How odd.

"You run the motor for a little while," he instructed. "I've got some work to do. Head for the neck of the bayou."

Curiosity's hooks were impaled deeply within me by now. I shoved Murph away from the bank and climbed aboard, casting furtive glances toward the bow as Herb leaned forward and began tinkering with the pole.

It was useless trying to sneak a peek at what he was doing. Whatever the source of his concentration, Herb was guarding it like a nuclear secret. Occasionally, I could hear the sound of metal blades jingling, but that's about all. He was still hunched over his mysterious treasure trove when we reached the long, S-shaped neck that winds out the back of the hollow.

I slowed the motor.

"Keep goin'," came muffled instructions from the front of the boat.

"I thought you always like to start here," I replied.

"You thought wrong. There's too many people around here today. They might see what I'm fixin' to show you."

"Too many people?!" I shouted. "Herb, how is anybody gonna see what you're doin'? There's not a soul within half a mile of us! We're surrounded by woods! We're in the wilderness!"

"No, *you're* the one who's in the wilderness," he said. "You're like those Israelites who fled outta Egypt and spent all that time wanderin' around lost. Fortunately, you've got me to show you the way. Take this boat around the next bend and stop at the stump on the left point, and I'll lead you right straight to the promised land."

I never cared much for the back end of Friendly Hollow, and Herb knew it. In the first place, the water there often transcends the word "stained." It's usually muddy enough, as Herb used to put it, to trail a raccoon without using dogs. In addition, the creek channel chokes down narrowly and is overshadowed by a never-ending series of tree limbs, making it all but impossible to cast.

But as I was about to discover, casting was not on the menu this day.

Herb squirmed halfway around in Pole Position and began extending sections of the fiberglass rod my way. It kept telescoping out, two to three feet at a time, until it was as long as the boat itself. Off the tip of it dangled one of Herb's sputter lures, complete with a feather tail.

"Off the tip" is the operative phrase here. The knot that held the lure in place was approximately four inches from the end of the pole. That's right: four *inches*. And even given Herb's propensity for strong line, I could see from my end of the boat that this was industrial-strength material. It wasn't even monofilament. It was braided cotton cord. With the combination of that pole and the rope, he could have derricked boulders off the bank.

"All these years, I've been showin' you how to wake bass up with a top-water bait," he said. "Today, we're gonna rock 'em back to sleep."

He grasped Mrs. Reed in one hand, dipped it into the water and paddled ever so slightly, just enough to position the front of the boat near the dark spot in the water above the stump. With his other hand, he held the butt end of the long pole.

He extended the tip until it was directly above the stump. He lowered the pole just enough that the lure touched the water. Then slowly, he began moving it—back-and-forth, back-and-forth, back-and-forth—in a figure-eight pattern.

"This is mesmerizin'," he whispered. "It lulls them ol' bass to sleep, just like those magicians do on TV when they dangle a pocket watch in front of somebody's face. Puts 'em in a trance so they don't know what they're doin'."

It does nothing of the kind. I've never spoken to a bass about the matter, but I daresay a fish is at the peak of alert efficiency when it attacks an object going "brrrr-brrrr-brrrr" above its lair. The bass may think it is a baby bird that has fallen into the muddy water. Or a crippled minnow. Or a frog. It doesn't matter. The bass has one instinct. Kill it and eat it.

Whoosh!

Brown water swirled at the tip of the pole, and Herb leaned back with all his strength. The creek channel literally exploded.

What followed was surely one of the most awkward bass-acquisition techniques ever performed. It was like watching a man try to pole vault inside of a drainage tile. Here we were, virtually swallowed up in a jungle, and Herb had a tiger by the tail at the end of a twelve-foot pole. Swinging the writhing largemouth into the boat would have been difficult had we been on open water. In this lattice of limbs, it was impossible.

But Herb had been there before, and he knew what to do. He shoved the butt end of the pole out the opposite side of the boat. In the blink of an eye, he ran his hands down the length of the shaft, fist-over-fist, until he reached the tip. And then, in one sweeping motion, he hauled the entire she-bang into the boat. He could not have been showered more thoroughly if he'd been seated in Murph, in the bed of my truck, and I drove through an automatic car wash.

"Now, *that's* the way you mesmerize a bass, Fingernail!" he said with a wide grin. "You just rock him to sleep and then—bang! —haul him in. If you'd be so kind as to pass me the stringer, I'll see if he has a big brother on up the creek."

"Let me get up there in Pole Position and try it a lick or two," I said.

"Not yet. You're still learnin'. I wanta make sure you have a good education before you start mesmerizin' one of these fish. Why, if I was t'send you into combat without a good education and somethin' awful happened, I could never forgive myself. *Herr-herr-herr-herr!*"

I threaded Herb's bass through a rope stringer, slipped the fish overboard, and ran the end of the cord through an oarlock. Up front, he was already applying Mrs. Reed in the direction of the next stump. As the boat neared the target, he set the paddle quietly at his feet and extended the pole once more.

"Brrrr-brrrr-brrrr," said the buzzbait as it turned in tight figure-eights.

"The secret is to not drop the tip of the rod in the water," Herb said. "If the tip goes in, it ruins the sound. Just keep it above the water, and keep it movin' in a figure-eight. Let the lure do all the talkin', just like wavin' that pocket watch. And 'fore you know it, a bass will be mesmerized."

The professor added another victim to the stringer before he deemed the student properly tutored. Then he sat back to critique. He did not have to search at length for points of discussion.

Until you try it, you have no idea how difficult it is to execute a tight, precise figure-eight pattern from twelve feet away. Especially when you are trying to (a) maintain a course with Mrs. Reed, (b) keep the bait swimming steadily, (c) not dip the end of the pole into the water, (d) weave the lure in and around overhanging limbs and vines, (e) brace yourself for a strike, and (f) endure a steady barrage of armchair quarterbacking from the opposite end of the boat.

After half an hour of fruitless exercise, the throbbing muscles in my arms begged for relief. The old pro and I changed seats. In approximately five minutes he brought Number Three aboard.

"A fellow who sleeps as much as you do oughta have this mesmerizin' down pat," he chided.

I persevered. And, sure enough, on my second tour of duty I managed to elicit a strike.

The water boiled. I set the hook. Then all of Herb's careful instructions went flying out the window. Instinctively, I tried to raise the pole. I couldn't help myself. In the heat of action, my mind went on autopilot. The business of hand-over-handing the pole never was considered.

I lost the fish.

Actually, I didn't so much "lose" the bass as I did "launch" it. I'm not exactly sure of the physics involved, but when a 190-pound man jerks down on one end of a twelve-foot-long pole with all his might, and a one-pound object is at the other end, the lighter object goes flying—even if the arc of the twelve-foot pole is somewhat impeded by tree limbs. The catapulted largemouth came crashing down two boat-lengths away and is probably still swimming in frenzied escape as I type these words.

Herb took a long pull from his water bottle and wiped his mouth with the back of his hand.

"It ain't as easy as it looks," he said, relishing his expertise.

But even dull students can learn a new trick over time. After a couple of sessions on the water, my mesmerizing skills improved dramatically. I never reached the pinnacle of the professor and probably never will. That twelve-foot pole was a virtual extension of his hand. He could drag figure-eights around stumps and across rocks as easily as doodling them on paper with a pencil. Yet by the end of that summer, he and I were lulling so many bass into slumber, it's a wonder the state health department didn't issue a warning about sleeping sickness.

Then a funny thing happened. We ended up mesmerizing ourselves.

It happened on a hot, hazy August afternoon at Herb's house. Just as we were about to pull away for a few hours of mesmerizing at Hot Dog Hollow, thunder rumbled in the west.

Only two things ever stood between Herb and his fishing. Snakes and thunder. He was mortally terrified of both.

Once, at the Golf Course, we had just electric-motored past a black snake sunning itself on a rock when the bass on the stringer beside Herb suddenly trashed in the water. He jerked violently and let out a scream like the devil himself had jabbed him with a pitchfork.

"What'n hell's wrong with you?" I hollered, trying to keep the Hub Tub from overturning.

"I thought that snake had me!" he exclaimed, white-eyed.

"That snake wouldn't hurt you if he was here in this boat."

"If that snake was in this boat, he'd be alone with you 'cause I'd be goin' out!"

"Herb, it's a black snake. They're harmless."

"You tell that to all them people who get snakebit every year," he replied with authority. "Happens all the time."

"Hardly anybody ever gets bit by a black snake. And even if they do, the snake won't hurt 'em."

"Baaaa-loney!" he snapped. "Any snake can hurt you."

It was useless to argue science with Herb—or politics, religion, economics, geography or any other topic, for that matter. But that little exercise on herpetology did open a wonderful window of opportunity for me. From that point on, anytime I felt Herb needed a jolt of excitement, I'd splash the water and shout, "Snake!" I trust he never told his cardiologist, or else the doc would have gone to court and gotten a restraining order against me. Aren't friends great?

But I did share his fear of thunder.

Thunder means lightning, and lightning can cook up a world of hurt for people in the outdoors. Snakes might not get my attention, but dark clouds and a low rumble in the distance certainly will. Thus, when the drums started booming that afternoon, we retreated to his house.

"It's probably just a scattered storm," I said. "Let's turn on the Weather Channel and see if it's on radar. Once it passes, we can go on out."

Herb settled into his easy chair and flicked the "on" button of the remote control. I plopped down in another soft chair on the opposite side of the room and slid a stool beneath my feet. As it turned out, we had just missed the local forecast. Another eight or ten minutes of commercials and updates on weather around the world would have to cycle before the local report returned.

It was cool and dark in his living room. The chairs were soft and inviting. Just like a sputter lure figure-eighting over a stump, the flickering image on the screen began to mesmerize us both. In less than a minute, Herb was snoring. Before the local forecast even reappeared, I was sawing logs, too. We slept over an hour before the telephone—from a telemarketer, of all things—bolted us both awake.

Naturally, it was all my fault.

"Here we are, supposed to be fishin', and you make me go back into this cool house, knowin' full well I'll drop off to sleep," he said. "You beat all, you know that?"

Now fully awake, we waited until the radar screen cycled back around. The storm, it revealed, had passed. On to Hot Dog Hollow we went.

Apparently the storm had rung the dinner bell down below. At almost every stump, point, tree top, and rock pile, we mesmerized a bass from hiding. We took turns sitting in Pole Position.

There is a string of small islands running along the upper end of Hot Dog Hollow. On the channel side of each island are scattered stumps. Not many. And with one notable exception, not large, either.

These stumps are not marked by any visible sign. They didn't need to be. Herb and I fished Hot Dog Hollow often enough to know their general location. With a bit of looking for a darkened shadow in the water, it was easy to pinpoint each stump. It got to where we didn't need to discuss orientation. The boat moved along, one stump to the next, as if it were on remote control.

It just so happened that I was sitting in Pole Position that afternoon, with the mesmerizing rod, as the largest stump loomed into view. As active as the bass had been since the passing of the storm, there was about a 99 percent chance a fish would be lurking alongside that stump. Probably a decent fish, too, because this was the largest stump in the entire hollow. The big dog lives wherever he wishes.

I knew it.

Herb knew it.

I kept sculling the boat, figure-eighting the buzzbait in silence. Herb began to grow nervous.

"Aren't your arms gettin' tired?" he finally spoke.

"Nope," I replied. "I'm fresh as a daisy."

In point of fact, my muscles were screaming for relief. Any other time, I would have gladly handed off the rod and taken a break. But not with pay dirt just up the creek.

"You're startin' to drop the tip of your pole," he volleyed. "Why don't you let me take it awhile?"

"Just give me a few more minutes."

Each of us knew what the other was thinking. Each of us would have bet ten dollars there'd be a bass on that stump. Each of us wanted a crack at it. Each of us would not dare tip our hands.

If you are clucking to yourself in righteous indignation right now—and thinking, "How utterly impolite of those two men! They call themselves friends, and yet they're trying to outmaneuver each other for the best fishing spot!"— you can forget about ever applying for membership in the Good Ol' Boy club because you simply don't understand the rules.

Herb and I were, and had been for years, the thickest of friends. It didn't matter one iota to either of us if the bass resting below that stump weighed eight ounces or eight pounds. If I had hooked it and he was the net man, he would have whooped and hollered just as enthusiastically as if the roles had been reversed. But there is a cardinal law of good ol' boyism that absolutely forbids yielding to the other until all options have been exhausted.

It's like the time my old friend Dan Culp was hunting quail with his father, V. L. Culp—or "Pappy," as everybody, except Dan, called him. Pappy was "Case" to Dan because, according to Culp family legend, he used to sing "Casey Jones" all the time as a boy.

In any event, the dogs had pointed a covey, and Pappy told Dan to walk in. Dan killed one on the rise.

Then the birds did a most unusual thing. Instead of flying into the most tangled cedar thicket on the farm as Ph.D. bobwhites are wont to do, they fanned out in the broom sage like imbecile quail from the nineteenth century. The dogs broke to begin vacuuming up the singles. That is when Pappy turned to the child he dearly loved, his own flesh and blood, the very fruit of his loins, and announced with the clearest of consciences, "Son, here's where the friendship ends."

"That damned Case!" Dan still laughs, over four decades after the fact. "He snap-shot every time before I could even get my gun to my shoulder. I never got to fire the first shell! Of all the wonderful times he and I had together, that's still one of my fondest memories."

Like I said—you either understand or you don't. If you do, welcome to the club. If you don't, perhaps there's a tea party you can attend.

So there Herb and I were, closing in on the Stump of Great Promise. You couldn't have budged me out of the front of that boat at gunpoint, despite the fact I was sore as a boil and begging for relief. And Herb, desperately wanting to take over the pole, knew full well that any hint of eagerness on his part would only encourage me to stay put.

The Stump of Great Promise grew closer and closer. Tar Baby didn't say nothin'.

Finally, Herb couldn't take it any more. He reached for his casting rod and shouted, "Duck!"

I ducked. Anybody would have. It was a reflex no human could have resisted. It's the same as when someone says, "Don't look now, but—" The first thing you're going to do is look.

It's a good thing I did bob my head downward a few inches, too. Otherwise, I would have been wearing a Jitterbug in my right jaw.

The plug barely cleared the top of my head. It landed softly four or five feet beyond the stump. Herb started it back toward the boat—*glub-glub-glub*.

"I just remembered somethin'," he said, steadily reeling and watching the path of his Jitterbug. "That stump's never been good for mesmerizin'. It always holds a bass that's already woke up and is hungry for a Jitterbug, like Ol' Redemption."

Of course it did. It was a nice bass, too. As it sloshed across the surface, Herb turned to me with a grin from ear to ear.

"Ooooh, I sure wish you could feel this fish! Man, is he fightin'! *Herr-herr-herr-herr!* But don't just sit there! Get the net! Get the net!"

To paraphrase Dan Culp: That damned Herb! Of all the wonderful times he and I had together, that's still one of my fondest memories.

The
Wailing Wall

With one notable exception, Friday, July 30, 1993, was a ho-hum day for news—one of those rare, calm moments in the newspaper business when assignment editors and graphics designers pace circles around their offices, sloshing lukewarm coffee all over the stained carpet and repeatedly asking each another, "You mean we've got nothing that warrants Page One?"

In Washington, work began on the Vietnam Women's Memorial. Gen. Colin Powell, chairman of the Joint Chiefs of Staff, officiated at the groundbreaking, noting that the new monument was "nine years in the making and more than twenty years in the needing." The Senate Judiciary Committee voted eighteen-to-nothing to recommend Ruth Bader Ginsburg's confirmation as the 107th justice of the United States Supreme Court. In national sports, the Atlanta Braves were seven games behind the San Francisco Giants in the western division of the National League. And on the local level, catcher Carlos Delgado hit his seventeenth home run of the season to help the Knoxville Smokies beat the Carolina Mudcats, five-to-one. Almost exactly seven years later, that same Carlos Delgado, now a first baseman for the Toronto Blue Jays, would hit a double in his first major league All-Star game, helping the American League to a six-to-three victory over the National League.

But it was the weather that grabbed most of the headlines and news space in July of 1993.

East Tennessee was withering under a relentless sun. In Knoxville, the paper reported that the 98-degree reading a day earlier had set a record for most consecutive days (thirty) with temperatures at or above 90.

In the Midwest, however, conditions were just the opposite. The Mississippi River and its tributaries were on a rampage. The same newspaper that told of parched conditions in east Tennessee described how the citizens of St. Louis, Missouri, were sandbagging feverishly, hoping to thwart the rapidly rising water lapping at their back door.

Herb and I experienced both of those worlds that day. We were surrounded by water and virtually helpless to do anything about it—and yet we were about to roast at the same time.

Things like that can happen when you go to the Wailing Wall.

"Wailing Wall" was Herb's code name for a concrete bridge, tunnel, and railroad trestle that bisects one arm of the Golf Course. You can quit looking for any more specific information. The Wailing Wall was then, and is now, one of the most prolific bass-producers in all of eastern Tennessee. It's as if a fisheries biologist had designed this spot in a laboratory.

There is a submerged rock pile beneath the right-hand, downstream wing wall, just after it comes off the bank. A flat section of the wall juts out from this rock pile, but only for a few feet. Then the bottom falls away. Bathed in the cool current flowing from the nearby tunnel, bass have over ten feet, vertically, to feed along thirty feet of the wall. It sounds hackneyed to say they stack up like cordwood in this area, but that's precisely what occurs.

Upstream, through the tunnel, the wing walls fan out to the shallows on both sides. The creek channel swings to the left-hand side, creating a similar drop-off like its opposite cousin through the tunnel downstream.

On the right-hand side, things get a bit dicier. Some of the rocks far down below are held in place with a wire gabion. If you so much as touch it with the hooks of a crankbait, you will donate your plug to Davy Jones. Worse yet, some of the wire on the gabion has eroded and broken loose through the years, creating even more snares and snags for lure donation.

Some of the time, you can jar your bait loose with a plug-knocker. Most of the time, you can't. I could make a sizeable down payment on a river bottom farm with the money I have invested on that hateful gabion.

Nonetheless, the place cranks out bass like school papers flying out of an old hand-operated mimeograph machine. Maybe once, over the decades that Herb and I visited this hallowed place, we failed to catch at least one bass. If so, I can't recall it.

Bass of all sizes, too. Mostly, they were healthy "chunks"—one- to two-pounders that seemingly lined up to waylay the next plastic worm, crankbait, or surface lure cast their way. When one was caught, another took its place.

The last time I competed in a bass tournament, I won it at the Wailing Wall. I hasten to point out this was the first, and only, time I ever won a bass tournament. I have sense enough to quit on top.

What made the victory truly special was the fact that I beat two of the best professional anglers in America—Bill Dance and Roland Martin. And before a rousing chorus of "Aw, what a bunch of horse hockey!" sweeps across the audience, let me stress this was not an honest-to-gosh, big-time bass tournament. It was more like a pro-am.

In the summer of 1982, as part of the World's Fair in Knoxville, the DuPont Corporation sponsored a nationwide search for the newest and strongest knot for its "Stren" line. A team of fishing's superstars—including Dance, Martin, and internationally known outdoor writers Mark Sosin, Jerry Gibbs, Ken Schultz, and Vic Dunaway—were brought to Knoxville to judge the entries.

As part of the festivities, DuPont hosted a four-hour bass tournament on Fort Loudoun Lake one morning. A number of east Tennessee's most heralded bass anglers were invited to participate alongside the pros. My name showed up on the short list not because of any remote hint of fishing skill, but for the same reason newspaper people frequently are asked to attend gala events: Somebody hopes they'll write about it.

Then I queered the deal by winning the damn thing.

The night before the tournament, there was a horrendous downpour. The runoff turned Fort Loudoun into a clay hole, headwaters to tailwaters. The bass shut down like someone had flipped a switch.

I was teamed with a DuPont official. I took him to the Wailing Wall. He caught nothing. I caught one, a largemouth weighing two pounds, eight ounces. Assuming the big boys had surely managed to catch their limits despite the terrible conditions, I almost tossed the bass back into the lake before returning for the weigh-in.

Thank heavens I didn't.

When we motored up to the tournament headquarters at Duncan Boat Dock, someone asked if we had anything to weigh.

I shook my head and said, "Naa."

No way was I going to embarrass myself by carrying a single, ho-hum bass up to the scales, certainly not in this esteemed group. Instead, I moored the boat, and my partner and I joined the crowd at the main tent.

Pretty soon, it was time to pack up and head to downtown Knoxville for the announcement of the knot contest winner. The tournament director got on a bullhorn and asked, laughingly and somewhat embarrassingly, "You mean *nobody's* got a fish to weigh?"

The place fell silent.

Then I said, "Well, I've got one little 'un."

And that—cross my heart and hope to die—is how I won a rod, reel, and several hundred dollars worth of fishing tackle. Turned out the pros had caught schools of bass, but none over the twelve-inch minimum size limit. As my boss, then-*News-Sentinel* sports editor Marvin West wrote in his column, "Venable insists it was a home-lake advantage and not because he was using another brand of line. Based on how little time Sam spends in the office, I think he practices more than the pros."

I vowed then and there never to compete in another bass tournament. It is a promise I have kept all these years. If I ever do break it—and may God smite me with a loathsome disease if that occurs—the Wailing Wall is the first place I will go if I absolutely, positively need something to take up space in a livewell.

But the Wailing Wall also was a regular producer of trophy-sized bass. As I write this chapter, I can look up at one of the exposed log beams in my office and see a mounted largemouth, seven-plus pounds, I caught there on June 20, 1991. With Herb's assistance, of course. Even though I have been lucky enough to take a few bass larger than it, I pledged the moment this one came into the Hub Tub that it would go to the taxidermist. When you catch one under the direct tutelage of the old master, it's special.

We had positioned the Hub Tub directly and exactly in the mouth of the tunnel that afternoon. "Directly" and "exactly" mean what they say because Herb had actually reached out and grasped the concrete wall with his hand and held the boat in place with his fingertips. A micrometer could not have positioned us more accurately.

"You know where that rock pile is, don't you?" he said.

"Yeah."

"OK. I turned a good bass right there the other day. I just had him on a couple of cranks, and then he came off. He's a decent fish."

I made a mental note to give Herb a thorough cussing later for not including me on that trip. I knew it could wait until we were finished with the Wailing Wall. No sense messing up pinpoint positioning with petty arguments. Especially when I was about to cast.

"He was layin' *juuust* off the rocks, near the wall," Herb coached. "If you cast too far out, which you'll probably do, you're gonna miss him. Put it right on the end of the rocks, and then come as close to the wall as you can get."

I was using a crawfish-colored Shad Rap. Unless there's wind, these things throw like a bullet. There was no wind. The plug landed softly in the 10-X ring of the bull's-eye. It submerged as I began rotating the handle of my reel. On perhaps the fourth revolution, the handle jolted to a stop. My graphite rod bowed.

"That's him!" Herb hollered as he reached for the net. "Now, don't horse him!"

What did Herb mean—don't horse him? As if I had some input on the matter, for Pete's sake? The fish was in complete control. The only thing I could do was hold on and hope Herb's constant sermons to me about sharp hooks and strong line paid dividends.

The bass tried to jump, but it was too heavy to completely clear the surface. So it bore into deep water. My sole contribution to the fray was to keep steady pressure on the rod. The big largemouth wore down rather quickly after that. It made one feeble run toward the tunnel, but Herb was quick with the mesh as it passed by.

"Fingernail, you can just call me your personal guide,"

he laughed, swinging the prize aboard. "I reckon since I've took you to raise, I've gotta show you ever'thing. *Herr-herr-herr-herr!* Now, you reckon you could hold this boat steady and let me get in a cast or two?"

On the sweltering hot afternoon of July 30, 1993—the day of no news—Herb brought a virtual twin to my bass into the Hub Tub. I did the netting honors this time. He and I hoo-haaed and high-fived and released the broad-shouldered warrior to fight another day.

Then Herb pointed the Hub Tub toward the back neck of the Golf Course. If we had known then what we were about to learn, we would have turned around immediately and headed for home.

What neither of us realized was that the prop on the electric motor was beginning to work loose. It managed to stay attached to the shaft until we were at the very end of the very last bend of the very farthest bayou on the entire Golf Course. Then it came off.

I was running the motor by this time. I reached for the controls to move us around a snag. All the motor did was hum. The Hub Tub remained stationary.

"How come you always wanta run so fast when we're usin' an outboard, but then you won't hardly move the boat at all when we're usin' an electric motor?" Herb asked.

I goosed the controls again. The boat didn't budge.

"I've got it cranked full blast," I said. "It must be hung on something."

I swiveled the motor around and pulled it out of the water. We both stared at the naked shaft, dumbstruck.

"I might have known you'd get me into a fix like this!" Herb finally blurted. "You got any idea how far up this bayou we are?"

"Three miles, easy," I answered. "Then we've got that long run past the Wailing Wall before we get back to the truck. I'll take Mrs. Reed for awhile, and then you can spell me."

It was at that moment Herb and I made a startling, cruel discovery. A paddle is fine for maneuvering an oval-shaped boat in and out of tight spots. But for direct propulsion, it isn't worth a dime. All it does is spin the boat in circles. We managed to travel about 150 yards, only 25 of it in a straight line, before I called a halt to the insanity. At this rate, it would be the middle of next week before we reached the truck.

Truthfully, I wasn't overly worried. It might take us until midnight, but I knew I could swim alongside the Hub Tub, pushing, pulling, wading, and dog-paddling, until I got us back to the ramp. But Herb had dialyzed that morning and was weak to begin with. An afternoon in the broiling sun had not improved his physical condition.

Herb removed the lid off of his foam cooler and used it alternately as a rudder and a paddle. Our progress was slow, but on a relatively straight course. We had been limping along for perhaps half an hour when suddenly, off the bank in front of us, an arc of monofilament shot out of the woods.

"Herb!" I exclaimed. "It's another fisherman!"

In all of the countless dozens of visits we had made to the Golf Course over the years, Herb and I had never encountered another soul. Occasionally we would see signs of bank fishermen—a bait cup here, a Y-shaped rod-holding stick there—but nothing with a human attached. And now here we were marooned, and what should appear but salvation on two legs!

It was a bitter pill for Herb to swallow. He was locked in intense internal debate over which was the lesser of two evils: Arduously paddling back to the truck by ourselves, or getting rescued by someone who—*arrgh!*—might discover his secrets?

"Don't say a thing about how we've been fishin' or what we've been catchin'," he hissed under his breath, just as the other angler came into view.

The guy on the bank looked like he'd come straight off the set of "Deliverance." Obviously, he hadn't seen other fishermen in "his" territory, either. Surely not two grown men in an over-sized bath tub. He stared blankly as we paddled toward him.

"You got a car around here anywhere, podnah?" I probed.

"Yeah," he answered, nodding to one side, "back over yonder at the road."

"You don't reckon you could give a man a ride back to the launch ramp on the main highway, could you?"

The guy didn't answer either way. All he did was ask, "You come from all the way over there to fish back here?"

"Yeah."

"You ever catch anything in here?"

Herb kneed me in the side of the leg. Hard.

"A catfish every now and then," I said with a shrug. "Sometimes a carp or two."

The pressure in the side of my leg relaxed.

"That's about all I ever catch, too," said the fellow whom, I suspected, was also attempting to market a whole-sale lie. "At least you get a little solitude."

We dragged the Hub Tub onto the bank, and Herb sat down with a plop. He was completely exhausted.

"Don't go nowhere," I told him as the rescuer and I hit the trail that snaked through the woods toward a distant road.

The trip around to the truck took fifteen or twenty minutes, during which time the guy and I did our dead-level best to match each other's lies. To hear us talk, the closest either of us had come to fresh fish in the last twenty-four months was the seafood section of a grocery store.

It was like two strange dogs sniffing each other out, except in reverse. Instead of growling our differences to one another, we continually groaned and lamented the fact that the lake was virtually fished out. Both of us agreed this was probably our last trip to the God-forsaken place.

The man dropped me off at my truck, followed me back to pick up Herb, helped us load up, and steadfastly refused our offer of gas money. No doubt he wanted to be done with us as soon as possible.

I pulled onto the gravel roadway and headed toward Herb's house. We hadn't gone more than a few hundred yards when the air conditioning lured him toward slumber.

Just as his head nodded, I said, "How many bass did that fellow have on his stringer?"

Herb roused awake.

"What makes you think I looked?" he answered indignantly.

"Because I know you. You looked at his fish and you checked out his tackle, too, to see if there was anything he might want to trade. I'll guarantee it."

"Well, I never heard to beat in all my life!" he snorted. "Here a fine feller stops his catfishin' long enough to rescue you'n me from the wilderness, and you go accusin' me of snoopin'. I wouldn't dare open another man's tackle box if I didn't know him. It's not polite."

"He wasn't catfishing, and you know it," I said with a laugh. "And no, you wouldn't open his box—but you sure as hell would take a peek at whatever was in plain view."

Herb grunted angrily. His face fell against the truck window as he resumed the snoozing position. Then his left eye opened in the narrowest of slits, and the hint of a smile flickered across his face.

"He only had a couple of little bass, 'bout like the size you catch," he said. "And his tackle wasn't nothin' but junk."

Rivers in
the Sky

I won't find out for sure until I pass through the Pearly Gates myownself—assuming all my transgressions have been erased from that window shade—but if any one of the faithful was ever able to take it with them when they left this orb, it was the Rev. Herbie.

Not gold, silver, or stock certificates, for sure. Aside from his lure collection, Herb wasn't affected by material things. He was a plain, simple man who loved his Lord, his family, his friends, and his fishing—more or less in that order.

Like any good preacher, he also knew his Bible. He could quote chapter and verse to address any element of human frailty, never wasting the opportunity to send a pointed message in my direction. As the infirmities of old age and ill health sapped more and more energy from his body, he turned to the Bible frequently for strength, courage, and guidance. I always accused him of cramming for the finals.

All mortals know their days on earth are numbered. Even though the blessed ones realize something better waits down the road, long after the shackles of humanity have been removed, they don't want to waste one moment of opportunity while they are still of the flesh. That's why Herb kept pushing himself to fish, even through his many bouts of pain.

Oh, he would falter temporarily. At the end of every fishing trip we made during the last two or three years of his life, he would insist this was his last. Absolutely no more.

"You have plumb wore me out this time," he would declare. "Don't call me again. If you take a notion to go fishin' in the next few days, fine. Go on without me. I ain't goin' no more. I hurt too much. Just leave me alone."

Invariably, my telephone would ring within seventy-two hours, and the imp would be right back up to his tricks.

"I reckon you've forgotten all about your ol' buddy," he'd begin.

"You told me not to bother you any more," I would reply.

"I said nothing of the kind! Why do you make up all those lies? I've been sittin' here, starin' at the phone, waitin' for it to ring. But I guess you've got more important things to do than fool around with an old shoe like me. Yes'ir, I'm just an old shoe, like you see on the side of the highway. You've done drove off and left me. I thought you were my friend, but my friend let me down."

I'd let him give me about twenty verbal lashes before asking, "When you want to go?"

"Go where?"

"You name it—the Golf Course, Watermelon Hollow, the Slop Hole, Pool Table Lake, Hot Dog Hollow. Whadaya feel up to?"

And before five minutes had elapsed, we would be deep into plans for the next journey—but only after I solemnly promised to bring him back the moment he grew tired.

Which I would do.

Which, he insisted, I didn't do the last time.

Which I would deny.

Which would lead to five minutes of no-you-didn't; yes-I-did, the likes of which is usually limited to kindergartens, courtrooms, and Congress. Then we'd hie off to the launch ramp. And after the trip, he'd vow he was never-ever-don't call-me-period-I-mean-it-this-time going again.

It was our version of perpetual motion.

Herb and I were keeping very few bass by this time. The concept of catch and release was sweeping through the ranks of bass anglers, but to be perfectly honest, that wasn't our motivating factor.

Herb was so weak and so dog-tired at the end of every outing, he didn't want to fool with cleaning a mess of fish. It wouldn't suit him for me to take them home, fillet them, freeze the meat, and bring it to him on our next trip, either. Freezing ruins fish flesh, he was convinced. According to his seasoned taste buds, a fish needed to either be eaten the day it was caught—one day later, at the outside—or put back into the lake.

A second reason that prompted us to liberate our bass had more ominous overtones. The state had begun issuing health advisories about consuming fish from many Tennessee waters, especially Fort Loudoun and Watts Bar. Even though a person would likely have to eat several pounds of fish a day—and keep up the pace for years—before suffering any adverse effect, the very notion that "something" might be wrong spared the life of many a mesmerized, crankbaited, and Jitterbugged victim.

It finally got to the point that Herb and I would keep spotted bass from Norris—I mean Sirron—and one or two sleek largemouths per trip from the slop holes. The fact that the slop holes smelled worse than a dog food factory didn't seem to dull Herb's love of fried fish. Nor mine. If the state decreed the slop holes were clean, who were we to argue?

In any event, numbers and poundage had long since ceased to matter. We simply enjoyed the act of fishing and each other's company. I once read a quote from Henry David Thoreau that sums up the feeling more accurately than anything I could say: "Many men go fishing all of their lives without knowing that it is not fish they are after."

Over time, Herb and I fell into a habit of what rural folks call "plunkin' around." We'd drive somewhere under the pretext of fishing and occasionally make a few casts. But just being outdoors, under the sun, on a riverbank or lakeshore, became an end unto itself. It is amazing how you start hearing, and appreciating, the buzz of insects and the cawing of crows when you are not concentrating on the sound of your top-water lure.

Herb also loved to plunk around on lake banks at low pool to look for lures other anglers had lost over the summer. He'd take a stick and start walking, flipping mud clods, rocks, and twigs with each step, until his eyes caught a faint glimmer in the sediment. Then he'd lean over and retrieve an absolutely worthless piece of rusted-hook junk that he "might be able to get a part off of." Into his pocket it would go.

Once, in the late fall on the banks of MacArthur, Herb spotted an interesting bird's nest in the limbs of a willow tree. Not so coincidentally, he had me along to climb the tree and wrest it down.

The nest was like none other I had ever seen. It was made almost entirely of old fishing line, along with strands of heavier twine. Herb couldn't have been any happier with a five-pound largemouth. He carefully carried it all the way back to my truck. J. B. Owen, the *News-Sentinel's* nature columnist, later identified it as the nest of a catbird.

"Those catbirds've got to have some Hubbard blood in them," I told J. B. "Hubbards and catbirds are the only two critters on earth who go ape over heaps of old string."

We also used fishing trips as an excuse to visit springs. I believe Herb knew the location of half the natural water outlets in this part of the state. On more than one occasion I drove him miles out of the way, just so he could drink deeply and then soak his bare feet in the frigid flow.

Such simple pleasures. Such bounteous rewards.

Yet there was one destination we often talked about at length but never visited. It was a piscatorial Land of Oz, way up yonder in the sky. Unlike Oz, however, this place truly exists. Herb said he had read about it in a magazine or newspaper—he couldn't recall which—but since he never produced the paperwork for proof, I was convinced he had made the whole thing up.

I was wrong.

After doing a bit of Internet searching, I came up with the article. It was credited to the American Geophysical Union and described research by a Massachusetts Institute of

Technology scientist named Reginald E. Newell—who gets my vote as Owner of Most Scholarly Sounding Name of the Century. How Herb came across such a highbrow article is anybody's guess. Truly, the Lord works in mysterious ways.

The article said the scientist made his discovery while analyzing data from satellites. According to Newell, the lower atmosphere of planet Earth is ringed by rivers of vapor, some carrying as much water as the Amazon. The entire system spans some 4,800 miles in length and is over 400 miles in width. It has waves and actually flows, just like a big river down below.

"On my way to Heaven," Herb used to say with a chuckle, "I'm gonna stop and fish in that thing. You know it's got to be unused water."

I teased him incessantly about his plans, pressing him for details about where he was going to launch his boat and what lures he planned on carrying.

"If it was me, I'd pack a variety. Top-waters, crankbaits, buzzbaits, maybe even live bait," I'd tell him. "Better take along plenty of grub, too. And gasoline. On a piece of water 400 miles wide, the marinas are probably few and far between. Better take something besides the Hub Tub, too. Those waves might be too much for it to handle."

Herb was not worried.

"You just leave ever'thing up to me and the Lord," he would say with a smile.

We laid the Rev. Herbie in the grave on March 28, 1999. It was a Sunday. That was the one day of the week he resolutely never fished.

But I'll bet you that bright and early the next morning, long before the heavenly hosts had begun to stir, Herb was backing his boat down the ramp, whistling merrily and singing "My Baby Thinks He's a Train." Free of pain forever.

And I'll double-dog guarantee he caught a bass.

For the
Record

In 1978, I was commissioned to write the history of Little Pecan Island, a three-mile-long ribbon of land, six feet above sea level, in the remote marshes of southwest Louisiana. Little Pecan was owned at the time by Herman Taylor Jr., an oilman and conservationist who lived in Natchitoches. It was the crown jewel of his Little Pecan Wildlife Management Area, an 11,000-acre spread teeming with ducks, alligators, deer, bass, crawfish, nutria, mink, otter, and other wildlife. Herman died of a heart attack in 1988, and his mecca has since changed hands several times, but my ten-year association with this wonderful land and its friendly people was truly a love affair.

During the course of research for my book, *An Island Unto Itself,* I discovered that the thin, shell-ridden soil of Little Pecan likely contains buried pirate treasure. That is, if buried pirate treasure actually exists anywhere along the Gulf Coast. Legends have a habit of taking on Herculean proportions as one century fades into the next, and here is no exception.

Without question, this region served as headquarters for pirates during the early 1800s. There is documented evidence that two of the most infamous of the lot—the Laffite brothers, Jean and Pierre—operated out of secluded camps near Little Pecan Island between 1815 and 1817. Since Little Pecan Lake offered fresh water for drinking and safety from coastal storms, it is within the realm of probability to assume the Laffites stored their booty on the island on occasion. Given the fact there is more profit to be realized by selling stolen merchandise than from hiding it, however, I have to think that most buried riches were exhumed and fenced rather promptly.

Nonetheless, the lure of treasure hunting is heady wine. Even today, many longtime residents of Cameron

Parish are convinced the proverbial pot o'gold still lies beneath the surface of Little Pecan Island, just waiting to be discovered. Every few years, someone happens upon a "long-lost treasure map" or hears word of a "death-bed confession" and investigates with shovel and metal detector. At last count, the most significant unearthing was a very old and very rusty wash tub.

Personal diaries, particularly of the hunting and fishing variety, are about as reliable.

Not that they aren't based on fact. Quite the contrary. These scripts are paragons of virtue—day-by-day recollections of what happened, when it happened, where it happened, and to whom it happened. Over the course of my career, I have been privileged to peruse diaries left in the wake of three legendary outdoorsmen: Dr. Joseph C. Howell, a University of Tennessee ornithologist and pioneer angler on TVA's earliest reservoirs; John T. Pirie Jr., a Chicago multimillionaire who hunted South Carolina quail at his leisure for more than half a century; and the Rev. Ray Hubbard, who neither earned a Ph.D. nor amassed a fortune, but left a cornucopia of entertainment in two handwritten volumes of memoirs.

As far as I'm concerned, the greatest value in any outdoor diary is the story it weaves. Passages from these journals are rough sketches, not detailed photographs. They reveal slices of life, not life itself. One cannot draw concrete conclusions about where to catch the biggest bass or find that secret covey of quail, for too many behind-the-scenes fragments of data often are omitted, too many variables fail to get recorded. These nuggets die with the authors, which is as it should be.

But if you want to get a feel for the life and times of Herb during our years together, look over my shoulder as I flip through the pages.

At the beginning, this was a nuts and bolts chronicle of bassing statistics. Herb's early entries listed the date of the trip (both the calendar date and day of the week), location, a brief statement about the weather, lures used, species caught, and size of the largest fish. Ever suspecting some sinister force might discover his secrets, of course, he frequently used codes. I daresay Herb's son Major, his grandson Patrick, and I are the only three people on this planet who know the precise body of water Herb was referring to when he penned such obscure notations as "The Rocks," "Pan Rock Lake," and "The Island."

Cryptography has its down side, however; if the formula gets too complicated, even the author becomes confused. That probably explains why, in the corner of one page, I found a set of translations that Herb had left for himself: "C-Bomb" for a worm-colored Bomber crankbait, "007" for a Rebel crankbait with red plastic flickers attached to the hooks, and "747" for his super-noisy sputter lure made from a Hula Popper body. I suppose that, like a double agent fluent in six languages, Herb occasionally needed a reference manual to remember that Hans in Munich was a friend, not an enemy.

Long before catch-and-release became a common practice, bass fishing was a game of compiling numbers. Weight was the unit of currency used to tabulate the success of any outing, and Herb was a money machine. I'll guarantee he is enjoying plenty of bass action in heaven because he sent enough big fish up there to last an eternity.

In the ten-year span from 1973 through 1983, for instance, Herb brought home 121 largemouth (and a handful of smallmouth) bass that ranged in weight from four pounds to nine-and-one-half pounds. Many were six pounds and heavier. In 1973 alone, he caught, cleaned, and presumably ate 245 bass—19 of which were trophy sized.

As cardiologists and ichthyologists would both agree, that's a lot of fried fillets. Many entries from Herb's later years include the word "released," much to the delight, I am certain, of his arteries and fisheries resources.

For several years in the mid-1970s, Herb entered a number of bass tournaments. He fared rather well in some of them, as witness his remarks from March 30, 1974— "Smoky Tournament, lunker prize, 5 pounds, 15 ounces." Almost exactly one year later, on March 29, 1975, he won $150 for a second-place finish with a total catch of 10 pounds, 11 ounces.

Not every contest paid off, however. On September 27, 1975, he simply said, "Fished tournament, no keepers, east wind, cold." And on April 23, 1977, he managed to produce only a single bass (one pound, ten ounces) in the Dogwood Arts Festival event. "Louisville was too muddy," he chided himself, "should have gone down the lake."

But fishing competitively against hundreds of other anglers was not in the man's blood, and by the mid-1980s he had given up tournaments altogether. He much preferred to go toe-to-toe with the angler in the other end of the boat. Especially if that other angler happened to be me. I chuckled aloud upon reading repeated entries where he boated five, six, seven or more bass during an outing and "Sam only caught one."

Indeed, as you surely have realized by now, Herb and I lived to tease one another. He blamed me for all mechanical failures—"With Sam to Cherokee, big motor wouldn't start," on June 2, 1987, for instance. And he insisted I was a wind magnet. On March trips to Douglas in 1994 and 1995, he noted, respectively, how I was responsible for "winds to 25 mph, boat almost capsyzed" [*sic*] and another blast that "blew Sam's cap into the lake."

Herewith a few other random examples of how Herb dearly loved to thrust his spear into my side:

"Three bass at The Rocks, with Sam. When Sam hooked one fish [I was using a plastic grub under a float for crappies at the time] he set the hook and his float hit me in the mouth," March 11, 1993.

"With Sam to Tellico, caught two trout for skins, Sam just took pictures," August 5, 1980.

"Ten good-sized bass, one channel cat, at Watermelon Hollow, caught on Gay Blade and Spot. Sam lost his Spot plug and boo-hooed all day long," October 11, 1994.

"With Sam to Blue Grass Lake, total of four nice bass, but Sam dropped one back in water trying to put on stringer," July 28, 1994.

"Fished on Pan Rock Lake with Sam. Caught seven bass on Dartin' R. Sam caught one," June 13, 1995.

Herb was lethal on local fish. But he seemed universally cursed during vacation fishing adventures. This is nothing new. As anyone who ever traveled to a far-off hotspot can attest, the dreaded phrase, "You should have been here last week" is liberally spoken by fish camp operators. "Started vacation," he wrote on April 7, 1975. "Went to Center Hill, water too high."

Conditions rarely improved. "To Woods Reservoir on company trip," he penned April 20, 1978. "Weather bad, stormy, hail, windy, cold."

Herb and Major journeyed to the Bass Galore Village on the Rainbow River in Florida from March 5 through 10 in 1977. "One month too early," he lamented on paper. "On Friday before we arrived, a 12-pound bass was caught near the dock. On Sunday, two 12s and a 14 were caught. Temperature was in the 80s, but dropped until we left. Went to Tifton, Georgia. Two weeks before, two were caught over 10 pounds."

Herb returned to Tifton in 1982 for a March 15-20 visit. His best bass weighed a mere five pounds. He ventured then to Allendale, S.C., then to Woods Reservoir in

Tennessee. All for a three-pounder. "Too early, water too cool," he noted. In May of 1986, he traveled with his wife Mary Jo to Lake Weiss in Alabama, the crappie capital of the South, only to discover "fish not being taken there." Perhaps he should have heeded his own admonition from May of 1978: "Due to weather, should have vacation in summer or fall, with one week in mid-March."

Even though Herb's journals are officially fishing diaries, they go far beyond the bounds of weights and measures. For example, I found three instances—July 24, 1990; June 23, 1991; and January 16, 1993—when he towed broken-down runabouts to safety with the diminutive Hub Tub. These gestures of nautical goodwill surely remained a source of great pride for the "tow-er" and one of deep embarrassment for the "tow-ee."

Indeed, there was room in Herb's outdoor memoirs for a wide variety of topics. Such as "mailed tax return" on March 9, 1992, and "Sam brought me some trick gifts and two dressed ducks" on January 8, 1993. As well as, "All fish from Cove Lake have been very good to eat if put on ice in a paper bag, kept in cooler, cleaned, and put back on ice until eaten," on May 20, 1994. Also, "Have caught three bass with T.W.R.A. [Tennessee Wildlife Resources Agency] tags," on August 29, 1981. And, "Caught six bass in Little River above Trundle's, one bass knocked a perch [bluegill] off the plug, and then got the plug." Plus two sad vocational notes from October 1988: "Had to discontinue preaching at Lost Creek Church due to health" and "made appointment for Social Security disability."

Another proclivity was Herb's habit of collecting printed material that included the name "Hubbard," especially if it related in some way to fishing. He was intrigued by a Texas reservoir, Lake Ray Hubbard. Once on a trip out west, he had his photograph snapped alongside a road sign advertising its location. His papers also included a flyer from the Orvis Company touting Hubbard's Yellowstone Lodge, an "executive fly-fishing retreat" in Montana.

I helped feed his unique habit by sending him "Hubbard-ized" printouts and clippings I happened to notice in the course of my work. Each piece of paper was inserted in his memoirs, along with whatever tongue-in-cheek comment I had scribbled at the time. The guy was such a packrat, he even kept the envelopes in which I had mailed my letters.

On February 12, 1992, for instance, the Associated Press moved a story out of Spokane, Washington, regarding one Charles R. Hubbard, who had been arrested for firing seven shots into a computer. "I bet the 'R' stands for 'Ray,'" I had written to him. Another AP dispatch, this time out of Nashville, reported the discovery of a dead patient at Meharry-Hubbard Hospital, sixteen days after she had turned up missing. Circling the references to "Hubbard," I noted, "Where else but here would this kind of thing happen?"

Once on a winter trip to Florida, I picked up a flyer offering guide services in the Kissimmee Chain of Lakes region. One of the experts listed was Harry Hubbard. "You can easily tell this outfit will rip people off and catch no fish whatsoever," I had joked in an accompanying note.

And reinforcing Herb's firm conviction that Hubbards throughout the world are cursed with bad luck, I sent a 1988 newspaper clipping detailing the agony experienced by a sixty-four-year-old woman who attempted to sail alone across the Atlantic Ocean. En route, her steering mechanism failed, as well as her automatic furling system. She had to haul in yards of drenched canvas when one sail toppled into the drink. "They don't have to tell anyone who this woman is," I wrote to him, circling Denise St. Aubyn Hubbard's name.

As if he had to be reminded.

Ever the beachcomber, Herb often would quit casting to poke along the exposed shoreline for lures other anglers had lost. Sometimes he discovered much more than fishing tackle.

"Found Indian head pennies at The Island," he inscribed on December 9, 1994. The following March 4, he was back at the Island and turned up "early relics, many saying 'U.S.' on them. Could this have been a fort at one time?"

He occasionally stumbled upon leftovers from law-breakers. Although I could find no reference to it in his diaries, I distinctly remember Herb's telling me about what he believed was a pipe bomb, found in the mud on Douglas Lake. He reported it to the police, but when he returned with authorities, the piece was gone. The notes do, however, indicate his December 15, 1994, discovery of an entire car, presumably stolen, sunk in the muddy bottom of Blue Grass Lake. It's probably still there, serving as the most unique fish attractor in this region.

Gardening was always on Herb's mind, too. Even if the casual reader did not know he had been raised on a farm, that fact would soon have become evident for there are constant references to crops. Nearly every year lists the first frost, rainfall and temperature patterns, and one of the most hallowed events in all of Tennessee agriculture, rural or suburban: the date of the first ripe tomato.

"Beans and potatoes are blooming," Herb recorded on May 18, 1988. "Had first radishes, planted March 12," he wrote May 10, 1989. On May 22 of the same year, the book says he "planted beets." Less than one month later, on June 17, he picked "first half-runner beans." However bountiful 1989 may have been, Herb must have soured on cabbage because in both June and July, he repeatedly reminded himself to not plant a crop the next year. "Plant only two rows of peas," he also chastised himself, along with "don't put the peas near the beans."

Not all crops came from his garden. "Dewberries ripe," Herb reported on June 12, 1989. And on September 23, 1993, he relished the gathering of "four gallons of musky-dines [*sic*] at Golf Course, nice size and color."

Although not trained as a scientist, Herb was nonetheless quite cognizant of the natural world. He made several references to the roe he found inside female bass after they should have spawned, such as this entry from August 15, 1983: "Stock Creek, five bass on Jitterbug, total eighteen pounds, still have eggs in them." On October 25, 1989, he found something entirely different in the bellies of two bass he caught from Little River: several large rocks. (Who knows? I wasn't along on that particular trip, but those two fish may have heard about my old trick of inserting pebbles to improve weight on the country store scales, and simply did the deed themselves.)

Also, on a number of occasions he left notes to himself about the comings and goings of Mother Nature's critters. "November 10 is when skunks hybernate [*sic*] for the winter," he observed in 1987. In 1990, he penned "hornets leave nest first of September." And on July 9, 1992, he listed an important milestone in the rites of summer: first katydid.

Few elements about climate and the seasons escaped his attention. Even the moon phases were recorded—in code, as usual. The full moon rated a "smiley" face, the new moon a "frowny" one. He also reported that, "September the wettest in 100 years," on October 3, 1977. And "Hurricane Hugo influencing weather," on September 22, 1989. And "record low temperature, 25 degrees," on April 12, 1989.

And "warm, windy" on February 22, 1977. And "Valentine surprise snow, six inches during p.m." on February 14, 1986.

That wasn't the only unique holiday listing. On December 25, 1982, Herb allotted himself a thirty-minute visit to the Rocks. Casting a black-and-gold Bagley crankbait, he unwrapped a superb Christmas present, a nine-pound largemouth.

A devout apostle of farming by the astrological signs listed in country almanacs, Herb also believed there was a similar correlation to fishing. Maybe there is. On May 6, 1976, the signs were in the heart when Herb extracted an eight-pound, four-ounce largemouth from a farm pond. On July 19 of the following year, the signs were in the bowels. Perhaps not a pleasant thought, but good enough for a seven-pound, four-ouncer from the Golf Course.

Football was another of his passions, particularly of the University of Tennessee variety, and he recorded the scores of many important games. Football was an interest we both shared, although Herb was, without question, the most pessimistic fan in Big Orange Country. Even if the Vols were favored to whip an opponent by five touchdowns, Herb would remain resolutely convinced they would go down in humiliating defeat until after the game was over. Then he'd start worrying about next week.

He is the only person I know—this is an iron-clad fact; ask Mary Jo Hubbard if you doubt my words—who got up from the television at halftime of the 1999 Fiesta Bowl, announced Tennessee was destined to lose to Florida State, and stormed off to bed. He was between the sheets, snoring like a foghorn, when Tennessee won the contest, twenty-three to sixteen, to claim the national championship. Herb didn't learn the victorious news until the following morning. He went to his grave regretting he ever told me about his gaffe, for I gigged him ceaselessly about it.

On the opening game of the 1996 season, UT athletic officials honored seventy-five years of football at Neyland Stadium by recognizing twenty-seven former players during the halftime festivities. Wearing his old number, one athlete from each three- or four-year period was to run onto the field through a T formed by the marching band.

Since some of the players from earlier campaigns were deceased, their sons were selected for the honor. My late father, Sam Venable Sr., a guard on the 1932 team, was picked for his era, and I was asked to wear his jersey.

When Athletic Department officials inquired who I'd like to have with me at the game, I gave them the names of my family—and one more: You Know Who. Herb and I sat side-by-side throughout the game. The only way I could have been any happier, or more proud, was if my own father had been there.

Herb's diary reflects his football interest, but with a unique twist only a fisherman could understand and appreciate.

On June 18, 1994, former Volunteer standout Heath Shuler, then a rookie quarterback for the Washington Redskins, appeared at a shopping mall in Knoxville to sign autographs as a benefit for a local hospital. Thousands of adoring fans showed up. Herb stood in line for six hours—7 A.M. until 1 P.M.—to get Shuler's signature. On an orange fishing lure.

"He said that was a first for him," notes the entry for that day.

On July 7, 1994, Herb was "making skin plugs for (head football coach) Phil Fulmer." On July 26, he "delivered five plugs to Coach Fulmer at UT." The effort must have resulted in a windfall of Fulmer fortune, for the afternoon log from that day exclaims "fished at Wailing Wall and caught about 15 bass."

Toward the end of his life, Herb all but ceased making diary entries. My guess is he was too exhausted at the end of the trip to do anything besides stumble into bed. What's more, his penmanship had grown almost illegible; it probably was too great an effort for him to even make notes. Most of his references by that time related to medical matters— "received two units of blood at UT Hospital," on October 11, 1997, for instance. Herb's last entry, on March 9, 1999, just sixteen days before his death, discussed his doctor's opinion that heart valve surgery was not an option because of the abundance of scar tissue around the organ.

Sad to say, there was no entry about our last fishing trip. We made it in early October 1998. By then, Herb was far too weak to endure the rigors of wading slop holes. I loaded him into my bass boat for a glorious, sun-splashed afternoon of casting on Douglas Lake. He slept almost the entire time. I caught a few bass and enjoyed the beautiful mountain scenery in the background. But there was an overriding pall of gloom. Not just the melancholy feeling one gets as autumn begins its slow slide into winter, either. I knew the sand had all but trickled out of his hourglass.

Oddly enough, I came home that evening with a most unique souvenir.

As I was removing the crankbait from the mouth of one especially frisky largemouth—you'd think I would have learned to use pliers by now—the fish shook violently and impaled my right index finger with the tip of one hook. The point didn't go in very far at all, not even to the barb. But it snapped off like a pencil lead.

The hateful sliver stayed inside, too. No matter how much digging I did with a sterilized needle over the ensuing weeks, I could not exhume this damnable irritant. And then, *voila*, as if by magic, the metal worked out on its own—just a few days before Herb died.

Which of his diary entries is my favorite? That's tough to say.

Untold dozens of them bring back pleasant memories. Herb's hen-scratch notations can make me chuckle almost as readily now as his *herr-herr-herr-herr* laughter did way back when. In my mind's eye, I can see him grinning devilishly as he formed the words on paper.

But in all those pages of wonderful moments, it's hard to top the simple notation he made on August 9, 1993: "With Sam to Wailing Wall, then to Golf Course, caught six bass, two three-pounders. Had a great time."

So did I.

Solo

Next to the drive I made to Miller Funeral Home to deliver the eulogy at Herb's funeral, the longest and loneliest journey I believe I ever took was to the Upper Slop Hole in September of 1999. In spite of the fact that it was a spectacularly beautiful day, crystal clear with a hint of autumn in the breeze, I was miserable.

Herb had been dead six months. I had reconciled myself to that. But I had not been able to bear so much as the thought of going fishing without him. Aside from a few hours of surf casting on summer vacation in Florida and one trout excursion to the mountains, I had not touched a rod and reel.

I knew there was only one way to get this terrible-tasting medicine down my throat. I needed to grab the medicine bottle and chug it and get back to the business of life. That's why I loaded my vest, hip boots, and casting rod into the bed of the pickup and headed up the highway.

This wasn't going to be easy, but I had to do it.

Grown men are not supposed to behave emotionally when a friend dies. Certainly—gasp!—not a male friend. We are conditioned to seal ourselves in a veneer of machismo and gamely plow on. "He was a good ol' boy, and I sure am gonna miss him," is about as plaintive an overture as we are allowed to make.

Been there. Done that. But not this time.

Driving along the roadway, alone, my mind fairly exploding with recollections of the many fun trips Herb and I had made on so many other fall afternoons, I couldn't help what happened.

First, a lump formed in my throat. Then tears welled up in my eyes. And before I could reach for a handkerchief to dab them away, they came cascading down my cheeks in hot, sobbing torrents.

I tried thinking of other things—like the work I had left back at the office. It did no good. About the time one side of my brain began concentrating on newspaper columning, the other side started spitting out Herbisms, and the waterworks started all over.

I tried thinking of the funny times—like the morning eight or nine years earlier, back when Herb was still hitting on all eight cylinders, and we got chased by a bull that suddenly appeared from over the hill as we waded into one particular slop hole. The diversion worked temporarily. I even chuckled aloud as the image of Herb slogging through the mud—frantically looking back over his shoulder one second and loudly blaming me for getting us into the mess the next—flashed across the screen of my mind.

The respite was short-lived, however. The tears burned on.

I remember thinking to myself, "Am I nuts to do this? Isn't this just like picking at a scab? Why not just turn the truck around and go back to work?"

And then just as quickly, I'd think, "Why, you big baby! Go on and fish! That's what you came up here to do, for Pete's sake! Quit blubbering!"

And on I would push.

Mercifully, the drive ended on an excellent note. When I pulled to the top of the hill overlooking the creek bottom, I could tell the conditions were optimum. Not only had the water dropped to a level that would give me nearly half a mile of fishing opportunities, I could already see swirls and splashes from feeding bass.

My psyche was virtually shouting by now.

"Now *there's* something to take your mind off moody thoughts! Get going!"

I parked the truck, squirmed into my hip boots and vest, grabbed the rod, and dropped off into the mudflat.

The place was as perfect—or as putrid, depending on personal preferences about mud—as I've ever seen it. For me, this could just as well have been the color cover of an outdoor magazine.

There was wind, but just barely enough to dimple the surface of the water and carry the musty smell of damp silt through the air. The bright sunshine was warm, but not hot. Overhead, a foursome of noisy crows ganged up on a screaming red-shouldered hawk. About halfway down the flat, a doe and two fawns—they'd probably scented me on the same breeze that carried mudflat perfume to my nostrils—crossed nervously from a grassy cove to a more secluded one with trees and brush. And way down at the end, where the creek pooled into the reservoir, sat a flock of perhaps two dozen blue-winged teal, surely the first of early autumn's migrants.

Things were even better from the bassing standpoint. This may well have been the same type of day Herb experienced when he first discovered the joys of slop holes.

It was like fishing in a hatchery. I could do nothing wrong. It didn't matter what lure I was casting. Whenever I'd see a swirl, I'd throw in front of it, make about two revolutions of the reel handle, and be fast into a leaping largemouth. I released most of them, keeping only a pair of two-pounders for the fillet knife back at home.

It was quite an odd experience. On the one hand, I was grieving the loss of a dear friend. On the other, I was enjoying some of the most spectacular, action-packed bass fishing in my life. It was like chewing down painfully on something hard during a seafood dinner—and discovering the culprit was a flawless pearl hidden among the oysters.

Then something extraordinarily strange happened. Something so eerie, it makes the hairs on the back of my neck stand up, even now.

In the second bend of this hole, where a tiny feeder creek comes in from the north side, there is a muddy pool perhaps twenty feet long and fifteen feet wide. A hump of rocks juts out of the mouth of the feeder creek and abruptly drops, eighteen or twenty inches into the hole. Bass are drawn to the base of this hump, probably because it creates a funnel for them to hem up schools of minnows. This was one of Herb's all-time favorite casting spots. I used to tease him about beating him to it as we climbed down the bank into the slop hole.

I would evermore beat him to it, obviously. Indeed, I worked the entire pool extensively that very afternoon and caught three or four small bass from it.

But just as I was about to move on down the creek bottom, I glanced back at the funnel. I don't know what caused me to do so. It was as if I could hear Herb shouting, "Hey, Fingernail! I guess you aren't gonna fish that place? Haven't I taught you anything? I'll swan, I don't know how you ever manage to catch a bass at all!"

There was a quarter-ounce, green and white Rattle Trap on the end of my line. I wheeled in my tracks, cast, and placed it—dead center, for once—at the mouth of the funnel, and began my retrieve. The rod bounced, then bowed, and then almost flew from my hands. The turbid water boiled angrily.

"Holy shit!" I whispered aloud, half expecting a sermonette about rocks and elephants to come wafting out of the clouds.

In water this shallow, "playing" a big bass isn't an option. You either winch him onto the bank, quickly. Or else the two of you part company, quickly. Approximately fifteen seconds, a few dozen reel cranks, and some interesting body English later, a fat largemouth—five pounds if it was an ounce—came skidding across the slick mud and lay flopping at my feet.

I should have released that bass. Really, I should have. But I didn't. This was a gift from Herb, just as surely as if he had knocked on my door, handed me the fillets inside a bag of crushed ice, and said, "I know your family can't depend on you to feed them, so I figured I'd do the job. *Herr-herr-herr-herr!*"

So I did exactly what the old master had taught me, what I know he would have done under similar circumstances. I added the big boy to the two juniors on my rope stringer. And then I took the stringer around the third bend, where nobody could see it—as if anybody else in their right mind would be out on this barren wasteland in the first place—and hid my fish in the next deep pool of the slop hole. Then I merrily continued fishing.

I worked my way back about an hour later. I had caught and released bass until the tips of both thumbs and forefingers were burred from their sandpaper teeth. I gathered the stringer and slopped back toward the truck.

I began climbing the rocks—going much slower than I did the afternoon I was certain Herb had suffered a heart attack and all he'd done was gone to sleep in the bed of my truck. At the top, I slid the three fish onto the ice in my cooler, opened a can of bellywash, and sat down on the tailgate to pull off my hip boots and soak in the last moments of the day.

There's not a lot of traffic on that road. Usually only one car or truck every half hour or so. But just as I was tying the laces of my shoes, a black car with a dent in the door pulled alongside and rolled to a slow stop in the gravel.

The driver didn't even step out. He just cranked down his window, stuck out his head, and asked, incredulously, "You been fishin' down there? In all that mud?"

"Yep," I replied, knowing full well that somewhere in heaven, an audible groan had just been issued.

"Catch anything?"

"Nope," I lied, safe in the knowledge that the heavenly groan had just been replaced by a smile of approval.

"What were you fishin' for in the first place?" he inquired.

I have no idea who that guy was. All I can tell you is that he had reddish hair and wore glasses, and that his black car with a dent in the door bore Sevier County plates. When he heard my reply, he drove off without saying another word.

Maybe he thought I was rude. Or a smart-ass. Or, given the fact that tears were welling up in my eyes once again, he might have assumed I was daft. It doesn't matter.

But I told him then, and I'm telling you now, that I spoke the truth, the whole truth, and nothing but the truth when I answered his question. He wanted to know what I was fishing for, and I gave him a full dose of the purest, cleanest, most undiluted honesty that ever crossed my lips, so help me Herb.

"Memories," I said.